T0276067

Managing Scientific Information
and Research Data

Managing Scientific Information and Research Data

Svetla Baykoucheva

ELSEVIER

AMSTERDAM • BOSTON • CAMBRIDGE • HEIDELBERG
LONDON • NEW YORK • OXFORD • PARIS • SAN DIEGO
SAN FRANCISCO • SINGAPORE • SYDNEY • TOKYO
Chandos Publishing is an imprint of Elsevier

CHANDOS
PUBLISHING

Chandos Publishing is an imprint of Elsevier
225 Wyman Street, Waltham, MA 02451, USA
Langford Lane, Kidlington, OX5 1GB, UK

Notices
Knowledge and best practice in this field are constantly changing. As new research and experience broaden our understanding, changes in research methods, professional practices, or medical treatment may become necessary.

Practitioners and researchers must always rely on their own experience and knowledge in evaluating and using any information, methods, compounds, or experiments described herein. In using such information or methods they should be mindful of their own safety and the safety of others, including parties for whom they have a professional responsibility.

To the fullest extent of the law, neither the Publisher nor the authors, contributors, or editors, assume any liability for any injury and/or damage to persons or property as a matter of products liability, negligence or otherwise, or from any use or operation of any methods, products, instructions, or ideas contained in the material herein.

British Library Cataloguing in Publication Data
A catalogue record for this book is available from the British Library

Library of Congress Control Number: 2015942342

ISBN 978-0-08-100195-0

For information on all Chandos Publishing
visit our website at http://store.elsevier.com/

  Working together
to grow libraries in
developing countries

www.elsevier.com • www.bookaid.org

Contents

Dedication   ix
Acknowledgements   xi

1 The road from chemistry—to microbiology—to information science   **1**
    References   **7**

2 Scientific communication in the digital age   **9**
  2.1   Introduction   **9**
  2.2   Challenging the traditional scientific publishing model   **9**
  2.3   The impact of the Open Access Movement on STEM publishing   **11**
  2.4   New models of scientific communication and publishing   **12**
  2.5   Use of social media in scientific communication   **15**
  2.6   Conclusion   **16**
    References   **17**

3 Ethics in scientific publishing   **19**
  3.1   Introduction   **19**
  3.2   Are we ever going to know the truth?   **19**
  3.3   Biases of editors   **20**
  3.4   Manipulating the impact factor of journals   **21**
  3.5   Peer-review issues   **21**
  3.6   Detecting scientific fraud   **22**
  3.7   How do researchers decide what to cite in their publications?   **24**
  3.8   Why do researchers resort to unethical behavior?   **25**
  3.9   Organizations involved in preventing unethical behavior   **25**
  3.10 Conclusion   **26**
    References   **27**

4 An editor's view: interview with John Fourkas   **29**

5 Finding and managing scientific information   **33**
  5.1   Introduction   **33**
  5.2   Discovery tools   **33**
  5.3   "Smart" tools for managing scientific information   **34**
  5.4   Information resources and filtering of information   **34**
  5.5   Comparing resources   **37**
  5.6   Conclusion   **40**
    References   **40**

6   Science information literacy and the role of academic librarians   43
    6.1   Is there a future for information literacy instruction?   43
    6.2   The many faces of information literacy   44
    6.3   Managing citations   45
    6.4   Designing information literacy instruction   48
    6.5   How do we know we are helping students learn?   48
    6.6   Assessing student learning   53
    6.7   Instruction formats   53
    6.8   Other elements of information literacy   58
    6.9   Sample questions for assignments in science courses   58
        References   62

7   Information literacy and social media: interview with Chérifa
**   Boukacem-Zeghmouri   65**
        References   69

8   Coping with "Big Data": eScience   71
    8.1   Introduction   71
    8.2   Types of research data   71
    8.3   Managing data   72
    8.4   Data standards   73
    8.5   Citing data   74
    8.6   Data sharing   75
    8.7   eScience/eResearch   76
    8.8   Data repositories and organizations involved in data preservation   76
    8.9   Data management plans   78
    8.10   eScience and academic libraries   79
    8.11   Conclusion   81
        References   82

9   Managing research data: electronic laboratory notebooks (ELNs)   85
    9.1   Introduction   85
    9.2   Recording research data   86
    9.3   Paper *vs* digital   86
    9.4   Finding information about ELNs   87
    9.5   Benefits of using ELNs   88
    9.6   Types of ELNs   88
    9.7   Introducing ELNs in academic institutions   91
    9.8   Conclusion   93
        References   94

10   The complexity of chemical information: interview
**   with Gary Wiggins   97**

11   Measuring academic impact   **103**
    11.1   Introduction   **103**
    11.2   The Institute for Scientific Information (ISI)   **103**
    11.3   The *Science Citation Index* (*SCI*)   **104**
    11.4   Journal Impact Factor (IF)   **104**
    11.5   *Journal Citation Reports* (*JCR*)   **104**
    11.6   The Journal Impact Factor is not without drawbacks   **105**
    11.7   *Essential Science Indicators* (*ESI*)   **105**
    11.8   h-Index   **107**
    11.9   Google Scholar Citations   **107**
    11.10   How do authors decide what and how to cite?   **108**
    11.11   More on evaluating journals   **109**
    11.12   Conclusion   **112**
           References   112

**12   From the Science Citation Index to the Journal Impact Factor
and Web of Science: interview with Eugene Garfield**   **115**
           References   **121**

**13   What it looked like to work at the Institute for Scientific
Information (ISI): interview with Bonnie Lawlor**   **123**
           References   **126**

14   Measuring attention: social media and altmetrics   **127**
    14.1   Introduction   **127**
    14.2   Measuring attention   **128**
    14.3   Altmetrics companies, applications, and tools   **128**
    14.4   Altmetrics and data provenance   **133**
    14.5   Conclusion   **134**
           References   **135**

15   Unique identifiers   **137**
    15.1   Introduction   **137**
    15.2   Unique author name identifiers   **137**
    15.3   Handling of author names by publishers   **140**
    15.4   Other unique identifiers   **143**
    15.5   Conclusion   **143**
           References   **143**

16   Epilogue: creating an information-literate generation of scientists   **145**

Index   **147**

For Sophie and Nicko

Acknowledgements

I feel greatly obliged to many people who have supported my research, shaped my thinking, and inspired me with their knowledge, vision, and life experiences. Those who had the most profound influence on my research career include Cécile and Jean Asselineau, Robert Brubaker, Eugene Garfield, Mayer Goren, Edgar Lederer, Guy Ourisson, Howard Sprecher, and Dimitar Veljanov.

The book would not have been what it is now without the interviews of Chérifa Boukacem-Zeghmouri, John Fourkas, Eugene Garfield, Bonnie Lawlor, and Gary Wiggins, who discussed scientific information, research data, and scientific publishing from different perspectives. Dr. Garfield also provided me with some unique images from his archive. They will remind readers how significant his contributions have been for us as users of scientific information.

I would like to thank George Knott, the editor of the book, Glyn Jones, the publisher, and Harriet Clayton, editorial project manager, for helping me with the manuscript and with all that was involved in publishing it. I feel greatly obliged to Donna Kirking, from Thomson Reuters, for her continuous support with EndNote and document formatting.

The interviews with the scientists, editors, publishers, and librarians, who graciously agreed to devote their time to answer my questions, are an addition to this book. My thanks go to all of them, including Alfred Bader, Grace Baysinger, René Deplanque, Michael Gordin, Morris Kates, Richard Kidd, Nigel Lees, David E. Lewis, David R. Lide, James L. Mullins, Maryadele O'Neil, Guy Ourisson, Maureen Rouhi, Eric Scerri, Leah Solla, Arnold Thackray, Bill Town, Andrea Twiss-Brooks, and Bryan Vickery.

During the eight years when I worked at the American Chemical Society (ACS) in Washington, DC, I was able to learn a lot about the chemical profession and scientific publishing, and I am always glad to see former colleagues at ACS national meetings.

I appreciate the support I have received from the University of Maryland College Park Libraries and particularly from Patricia Steele, Dean of Libraries, and Gary White, Associate Dean for Public Services. I benefited from the discussions I had with Jeremy Garritano, Karl Nielsen, Terry Owens, Carlen Ruschoff, and Ben Wallberg, who shared with me their expertise in such areas as discovery tools, eScience, open access, institutional repositories, and research data resources. Alla Balannik and Peter Armstrong, staff members in the White Memorial Chemistry Library, deserve my special thanks for absorbing some of my administrative responsibilities while I was writing this book.

It was a big honor for me to receive the prestigious Val Metanomski Meritorious Service Award from the Division of Chemical Information (CINF) of the American Chemical Society. The award is given to members who have made outstanding contributions to the Division, and it recognized my contributions as editor of the *Chemical Information Bulletin*, in which capacity I served for five years. I also highly appreciate the support CINF leadership have given for this book in allowing me to include substantial quotes from interviews that I have published in the *Bulletin*.

The book includes many images from databases and I would like to thank the following publishers for allowing me to use their images: the Chemical Abstracts Service (a division of the American Chemical Society), Altmetric, Elsevier, Impactstory, LabArchives, Plum Analytics, Springshare, SurveyMonkey, and Thomson Reuters.

I appreciate the significant sacrifices that my family has made in supporting me through my career. My parents and my sister provided me with a social environment that stimulated my interests in languages, history, literature, and philosophy. My daughter Vess has accepted the fact that my attention to her was always shared with preoccupation about my work or other interests. The biggest sacrifice was made by my husband Simeon, who has given me unselfish support throughout my whole career. As the first reader, honest critic, and careful copy editor of everything I was writing, he moderated my thoughts and prevented me from going to extremes in opinions and lengthy explanations.

The road from chemistry—to microbiology—to information science

1

Hang on to ideas until they mature.

John Mellencamp (on The Today Show, *September, 2014)*

In the technologically complex environment in which we are working today, the fear of missing some important piece of information is becoming more and more palpable. Writing a book is an endless process, especially if the topics you are writing about are changing very quickly. You think you have included the latest information and then suddenly you learn of something you have never heard of before that is common knowledge to others. At some point, you have to stop gathering information, reading articles, discussing your book with people, attending conferences and taking notes, and reading e-mails sent to Listservs.

People write books for different reasons. Sometimes they even cannot explain why they are doing it. As for me, I know why I wrote this book—it is because of Eugene Garfield. Early in my research career, I became fascinated with his essays published in a little weekly journal called *Current Contents* (Garfield, 1979). These essays triggered my interest in information science (Garfield, 2014) and the thought that someday I would be part of this world has always stayed in the back of my mind and later made my transition from the lab bench to librarianship seamless.

Eugene Garfield created the Science Citation Index (SCI) and was the founder of the Institute for Scientific Information (ISI) in Philadelphia. The SCI became the basis for important information products such as Web of Science, Essential Science Indicators, and *Journal Citation Reports (JCR)*. Several years ago, I interviewed Dr. Garfield (Baykoucheva, 2006) and later regretted about not asking him some more questions. I feel honored that he agreed to do another interview for this book, which is included in Chapter 12.

It is sometimes difficult for people with nontraditional careers to explain to others what has driven them through all their professional paths. The scientific revolution that has led to major discoveries in science and technology—the advancements in the exploration of space, the elucidation of the structure of DNA, and the discovery of new drugs—created an atmosphere of optimism about the role that science could play in making society better, and I wanted to be part of this revolution. The life and work of Marie Curie played a significant role in my choosing chemistry for my undergraduate education. I was prepared to endure the hardships of a scientific career—spending endless hours in the lab while working with toxic chemicals and infectious agents—to live out the experience of being a scientist.

My interest in understanding how microorganisms cause diseases and how these diseases can be prevented or cured made me redirect my career from chemistry to infectious microbiology. I spent a significant part of my research career at the Institute of Microbiology of the Bulgarian Academy of Sciences. My initial research was focused on the chemical basis of bacterial pathogenicity and the mechanisms by which virulent strains of bacteria survive and overcome the defense systems of the body. It was not easy for me, being trained as a chemist, to learn how to work in sterile conditions and protect myself and my colleagues from getting infected with the dangerous bacteria we had in the lab. When I started my graduate work, my daughter was ten months old, and I still remember how every night, before going to sleep, I tried to remember all the steps and procedures that I had gone through in the lab during the day, to make sure that I had not exposed myself to the highly virulent strains of bacteria that I was handling.

Along with performing research in the lab, I began writing articles for popular science and literary journals on a broad range of topics. My interest in languages created a parallel career for me as a translator and editor of scientific and other publications. The long list of my activities as a translator includes a psychology book on transactional analysis (*I'm OK, You're OK* by Thomas Harris, originally published 1969), which was published by a major publisher in Bulgaria.

My stay in Paris for one year as a postdoctoral fellow of the International Atomic Energy Agency opened a new chapter in my life, as it allowed me to learn new modern research techniques and broadened my interests in European culture, languages, and history. Upon my return to Bulgaria, I wrote articles about different aspects of cultural life in France, historical places that I had visited, and books that I had enjoyed. An article about a literary TV talk show, *Apostrophes*, which had dominated French intellectual life and had influenced the reading habits of the nation for more than a decade, was published in a popular weekly newspaper with very large circulation.

An essay about Zelda Fitzgerald's book, *Save Me the Waltz*, published in a literary newspaper, reflected my interest in the life and works of American writers living in Paris in the 1920s—an interest I have preserved until today. How could I have imagined that this essay, written in a Slavic language, would be downloaded over 100 times from a US university repository 30 years later?

In 1990, after applying for a position for a visiting scientist advertised in the journal *Science*, I came to the United States and worked at the Department of Microbiology, Molecular Genetics, and Immunology of the University of Kansas Medical Center in Kansas City, Kansas. There, I investigated the effects of the cellular membrane lipid environment of macrophages on both the expression of cell receptors for bacterial lipopolysaccharides (LPS) and the capacity of these cells to respond to LPS by producing tumor necrosis factor (TNF).

The most productive period in my career as a scientist came when I was offered a position at the Department of Medical Biochemistry of Ohio State University (OSU), where I had the opportunity to collaborate with Dr. Howard Sprecher, a distinguished scientist in the field of fatty acid metabolism. Our studies led to discovering new fundamental information about the way fatty acids are synthesized and degraded in the liver by small membranes called peroxisomes. These findings provided an important clue to how the brain and the eyes obtain specific fatty acids required for normal organ function. The nine papers that we published in some of the most prestigious biochemistry journals continue to be cited every year.

A career change is a serious thing and has to be prepared years before making the decisive step. Reading Eugene Garfield's essays for years and publishing articles on topics outside of my area of research made this transition easy for me. While still doing research, I enrolled in Kent State University's master's program in library science, which had a branch on OSU campus. The beginning of my library career coincided with the exponential growth of the Internet, which provided me with many job opportunities both in academia and in special libraries. For eight years, I was manager of the Library and Information Center (LIC) of the American Chemical Society (ACS) in Washington, DC, a position that entailed providing information to the editors of the ACS journals and particularly to the editors of the ACS flagship weekly magazine *Chemical & Engineering News* (*C&EN*). For an important publication with a large circulation (around 200,000), having accurate content was vital, which, combined with the short deadlines, put enormous pressure on me. For five years, I also served as a voluntary editor of the ACS organizational monthly newsletter *The Phoenix*. At the ACS, I was able to gain an insider's view of the scientific publishing field, attend many professional conferences in the United States and abroad, and establish long-lasting connections with many scientists, editors of scientific journals, publishers, and librarians.

In 2005, I joined the University of Maryland (UMD) Libraries in College Park as head of the White Memorial Chemistry Library, where I am responsible for the day-to-day management of a busy branch library (with annual attendance of around 145,000), serving also as a subject liaison for chemistry and biochemistry, teaching scientific information, performing collection development, and doing research. Teaching is an activity that has given me a lot of satisfaction. In the past ten years, I have conducted over 300 library instruction sessions (over 6000 participants) in a broad range of undergraduate and graduate courses, as well as in Professional Writing Program courses and honors programs.

In 2010, together with two other librarians from UMD, I took part in the eScience Institute, a six-month educational course offered by the Association of Research Libraries (ARL). As part of this program, we evaluated the readiness of the university to support eScience, interviewed administrators, gathered information about similar initiatives in peer institutions, and wrote a report, which proposed steps for implementing eScience support on campus.

As readers will see, a major theme in this book is STEM publishing. STEM stands for "Science, Technology, Engineering, and Mathematics." It is an inclusive term used to separate these disciplines from other areas of scholarship. STEM is usually part of a compound name—STEM disciplines, STEM departments (in universities), STEM education, and STEM publishing. Governments have started paying more attention to STEM (education and publishing, in particular), and they are concerned that universities are not up to the task of preparing students for the new challenges imposed by new digital technologies and global competition. So STEM, in whatever context this term is used, is "en vogue."

Organizing scientific information is at the core of doing science. We cannot imagine what science would have looked like today without the Periodic Table of the Elements in which Dmitrii Mendeleev not only arranged the existing chemical elements but also included reserved spaces for those not yet discovered (Gordin, 2004; Scerri, 2006). The management of scientific information starts with how scientists gather information, organize their data, and communicate their findings. Today, they can "hang out" in the same environment where they can do so many things—search for literature and property information at the same time; see how many times an article they were looking at has been viewed, downloaded, and cited; forward an interesting article to others and comment on it; and find out what others are saying about their own research on Twitter and Facebook. Creating, organizing, searching, finding, and managing scientific information are all "moments" that blend seamlessly with research activity at the lab bench and into our lives.

With science becoming more and more interdisciplinary and the volume of data growing at unprecedented speed, there is a need to look at scientific information and how we manage it from a new perspective. The digital environment and interactive technologies using Web 2.0-based tools allow performing research and organizing, managing, and sharing scientific information in a much more efficient way.

As editor of the *Chemical Information Bulletin*, published by the Division of Chemical Information (CINF) of the ACS, I had the opportunity to interview librarians, researchers, editors, publishers, and experts in scientific information who shared their views on scientific information and publishing. The preliminary work that I did in preparation for these interviews and the discussions that followed have informed my thinking, writing, and teaching. I am very grateful to all who have generously shared their knowledge and enthusiasm about the new developments in their particular fields. I consider these interviews to be an integral part of this book—they are an extension to it. All interviews can be viewed at the CINF Web site at www.acscinf. org/content/interviews, but I would like to mention here those of them that are directly related to the topic of this book.

In 2007, Bryan Vickery, who was then editorial director of Chemistry Central, discussed the open-access movement and how it was affecting STEM publishing (Baykoucheva, 2007b). Maureen Rouhi, editor in chief of the ACS magazine *Chemical & Engineering News*, analyzed the challenges presented by the globalization of science and how it affects scientific publishing (Baykoucheva, 2013).

Alfred Bader, founder of the largest company for research chemicals in the world, the Aldrich Chemical Company (now Sigma-Aldrich), who is also a scholar of the history of chemistry and a renowned expert on Dutch paintings, shared with me his enthusiasm about the connections between chemistry and art and deliberated on the authorship in scientific discoveries (Baykoucheva, 2007a).

The impact of Russian chemists on European science and how one of the most remarkable resources for chemical information, the *Beilstein Handbook of Organic Chemistry*, was created are discussed in an interview with David E. Lewis, chemistry professor at the University of Wisconsin-Eau Claire (Baykoucheva, 2008b). The chemistry database we now use under the name of Reaxys includes the electronic version of the *Handbook*. Arnold Thackray, a former president of the Chemical Heritage

Foundation in Philadelphia, looked at the history of this organization and its role in promoting chemistry and chemical information (Baykoucheva, 2008a).

The *CRC Handbook of Chemistry and Physics* is a major chemical information resource. David R. Lide, its long-time editor in chief, describes in his interview how such a complicated handbook has organized chemical information in a unique and easy-to-use way (Baykoucheva, 2009). In her interview, Maryadele O'Neil, senior editor of *The Merck Index*, described how this versatile chemical reference resource was compiled and published (Baykoucheva, 2010b). *The Merck Index* was acquired by the Royal Society of Chemistry in 2012.

Eric Scerri, a philosopher of science at UCLA, looked at scientific discovery, organization of information, and the significance of the Periodic Table of the Elements from a historical and philosophical point of view (Baykoucheva, 2010c). Michael Gordin, director of Graduate Studies in History of Science at Princeton University, showed how the cultural environment "of an epoch, a country, a region, or an organization influence the developments in science and the public attitude about it" (Baykoucheva, 2011b). He pointed out to the importance of such factors as "the experimental equipment and resources available to the scientist, his or her level of education and preparation, access to communication from other scientists, and the general state of science at the time." Gordin also discussed plagiarism, scientific fraud, and pseudoscience from the point of view of a historian of science.

James L. Mullins, Dean of Purdue University Libraries, discussed eScience and the involvement of academic librarians in supporting data management (Baykoucheva, 2011c). The challenges for science librarians in a "big data" world were discussed in interviews with Grace Baysinger, the chemistry librarian of Stanford University (Baykoucheva, 2011a), and Andrea Twiss-Brooks, Co-director of the Science Libraries Division of the University of Chicago's John Crerar Library (Baykoucheva and Twiss-Brooks, 2012).

This book discusses topics and issues that are both broad and complex. Librarians will find in it information about new areas in which they could get involved to support education and research in their institutions. They can also use some of the presented practical solutions when teaching scientific information and citation management. Vendors of scientific databases may find some useful feedback about how their products are used and what needs to be improved.

How scientific research is evaluated, recognized, and perceived affects the lives of scientists, the distribution of research funds, and the entire process of doing science. The book provides information about metrics to evaluate the quality of research and how the new area of altmetrics is trying to measure attention to research. Readers will learn what eScience is and how research data can be managed by adopting electronic laboratory notebooks. The strategies for managing scientific information and research data suggested in the book could be of interest to students, researchers, and librarians and to those who just want to make their information gathering more efficient.

To many people, "managing scientific information" means using EndNote, Mendeley, Zotero, or other bibliographic management programs. But it is more than that—it is a process that depends on how scientists communicate their findings and opinion; how their articles are evaluated, published, and disseminated; how secondary publishers and social media cover, filter, and make scientific information available; and how readers access this information, share it with others, and use it. New technologies, new thinking, and social networking are changing how research is performed and communicated today. The 16 chapters of this book look at how scientific information and research data can be made more manageable:

Chapter 1 is an introduction to this book. It explains why the book was written and how the topic of managing scientific information and research data could be viewed from different angles and perspectives.

Chapter 2 discusses the changes in scientific communication and STEM publishing. As the expectations of making research results openly available are growing and traditional forms of peer review and publishing are challenged by emerging models, researchers are changing how they communicate their findings.

Chapter 3 deliberates on the consequences of scientific fraud and plagiarism and the impact of unethical conduct on the careers of scientists and the reputations of institutions. This chapter also includes information about some new technologies that scientific publishers are using to detect manipulation of data and plagiarism before the papers submitted for publication enter the publication process.

Chapter 4 is an interview with the associate editor of the ACS *Journal of Physical Chemistry*, who gives an insider's view on how articles submitted for publication are processed and evaluated.

Chapter 5 could be looked at as a guide for searching, refining, and managing scientific information.

Chapter 6 compares different bibliographic management programs and outlines strategies for integrating bibliographic management in the teaching of scientific information.

Chapter 7 includes an interview with Chérifa Boukacem-Zeghmouri, lecturer in Information and Communication Science at Claude Bernard Lyon 1 University, who gives an interesting perspective on how graduate students and experienced researchers in French academic institutions gather information and use social media.

Chapter 8 discusses what eScience is about and how academic libraries are getting involved in supporting the data management needs of researchers.

Chapter 9 presents an overview of electronic laboratory notebooks (ELNs) and how they can be used to record, share, manage, and preserve research data.

Chapter 10 includes an interview with Gary Wiggins, one of the most prominent figures in the field of chemical information, who discusses the challenges presented by the complexity of chemical information and the changing role of science librarians.

Chapter 11 is a critical analysis of metrics such as Journal Impact Factor (IF), h-index, Google Scholar Citations, and Scopus Journal Analyzer, which are used to evaluate scientific research.

Chapter 12 includes an interview with Eugene Garfield, who describes how he came up with the idea of using citations included in scientific articles to organize and manage information. He also provides a historical perspective of the *Science Citation Index* and how it has "underwritten" what we now know as Web of Science.

Chapter 13 includes excerpts from a previously published interview with Bonnie Lawlor (Baykoucheva, 2010a), who worked at the Institute for Scientific Information (ISI) for 28 years and later held high-level positions at other organizations dealing with scientific information. In her interview she vividly describes the atmosphere at ISI during those pivotal years when such innovative products as the *Science Citation Index, Current Contents*, and other science resources were created and what it was like to work with Eugene Garfield.

Chapter 14 looks at the use of social media by researchers and how the new area of altmetrics is attempting to measure attention to research.

Chapter 15 discusses the unique authors' identifiers, such as ORCID, ISNI, and ResearcherID and how publishers are handling authors' names. Information about some unique identifiers for chemical names is also included in this chapter.

Chapter 16 captures the author's views about the future involvement of academic librarians in supporting research and education in their institutions.

At the 2012 annual conference of the International Federation of Library Associations and Institutions (IFLA) held in Helsinki (Finland), a discussion took place during one of the presentations, when a librarian from a Finnish public library reported that their library had introduced two new services. They called one of these services "Ask Us Anything," and the other one was promoted as "Ask a Librarian." While the first service had been bombarded with requests, the "Ask a Librarian" barely received any. They concluded that using the word "librarian" might have discouraged people from using the service, as users might have perceived it as a more formal engagement. Academic libraries need to reimagine themselves and find new roles to play in their institutions. What has worked in the past is not working anymore, and librarians are already feeling the impact. The consequences of not doing it soon enough could be very significant. In this book the challenges for academic libraries posed by the new technologies and higher expectations from higher education are discussed along with the new roles they could play in the future in supporting research and education in their institutions.

We should really feel lucky that we are living at a time when so much scientific information is available and so many sophisticated tools allow us to retrieve, refine, and manage it. We also need to realize that there is a lot of incorrect and incomplete information on the Internet, sometimes even in journal articles that have been vetted through peer review. As every scientist knows, the provenance of data is a prerequisite of whether we will trust the information or ignore it. I would like to suggest that even if you trust what you hear and see, still verify it.

Looking back, I see how fortunate I have been to come across so many interesting opportunities and meet such extraordinary people. The scope of my research has allowed me to establish close professional and even personal ties with many scientists in the United States, France, and many European countries. In this book, I have shared my own experiences and showed how modern technologies and the Internet have changed how science is performed, communicated, and shared. I hope I have succeeded in this effort.

References

Baykoucheva, S., 2006. Interview with Eugene Garfield. *Chem. Inf. Bull.* 58 (2), 7–9. http://acscinf.org/content/interview-eugene-garfield.

Baykoucheva, S., 2007a. Chemistry and art: the incredible life story of Dr. Alfred Bader. *Chem. Inf. Bull.* 59 (1), 11–12. http://acscinf.org/content/chemistry-and-art-incredible-life-story-dr-alfred-bader.

Baykoucheva, S., 2007b. Paving the road to more open access for chemistry: interview with Bryan Vickery, Editorial Director of Chemistry Central. *Chem. Inf. Bull.* 59 (1), 13–15. http://acscinf.org/content/paving-road-more-open-access-chemistry.

Baykoucheva, S., 2008a. The Chemical Heritage Foundation-past, present, and future: interview with Arnold Thackray. *Chem. Inf. Bull.* 60 (2), 10–13. http://acscinf.org/content/chemical-heritage-foundation-past-present-and-future.

Baykoucheva, S., 2008b. The Russian invasion … in chemistry: interview with David Lewis. *Chem. Inf. Bull.* 60 (1), 7–10. http://www.acscinf.org/PDF/lewis2008.pdf.

Baykoucheva, S., 2009. The CRC Handbook of Chemistry and Physics: a mountain, a cathedral, a battleship, or … an iceberg? An interview with David R. Lide, Editor-in-Chief. *Chem. Inf. Bull.* 61 (2), 4–8. http://www.acscinf.org/content/crc-handbook-chemistry-and-physics-mountain-cathedral-battleship-or-iceberg.

Baykoucheva, S., 2010a. From the Institute for Scientific Information (ISI) to the National Federation of Advanced Information Services (NFAIS): interview with Bonnie Lawlor. *Chem. Inf. Bull.* 62 (1), 17–23. http://acscinf.org/content/institute-scientific-information-isi-national-federation-advanced-information-services-nfais.

Baykoucheva, S., 2010b. The Merck Index, an encyclopedia of chemicals and natural products: interview with Maryadele O'Neil. *Chem. Inf. Bull.* 62 (3), 5–9. http://acscinf.org/content/merck-index-encyclopedia-chemicals-and-natural-products.

Baykoucheva, S., 2010c. A philosopher's view on the Periodic Table of the Elements and its significance: interview with Eric Scerri. *Chem. Inf. Bull.* 62 (1), 27–32. http://www.acscinf.org/content/philosophers-view-periodic-table-elements-and-its-significance.

Baykoucheva, S., 2011a. A new reality for academic chemistry librarians: an interview with Grace Baysinger. *Chem. Inf. Bull.* 63 (3). http://bulletin.acscinf.org/node/211.

Baykoucheva, S., 2011b. Political, cultural, and technological impacts on chemistry. An interview with Michael Gordin, Director of Graduate Studies of the Program in the History of Science, Princeton University. *Chem. Inf. Bull.* 63 (1), 50–6. http://www.acscinf.org/content/political-cultural-and-technological-impacts-chemistry.

Baykoucheva, S., 2011c. What do libraries have to do with e-Science? An interview with James L. Mullins, Dean of Purdue University Libraries. *Chem. Inf. Bull.* 63 (1), 45–9. http://acscinf.org/content/what-do-libraries-have-do-e-science.

Baykoucheva, S., 2013. Interview with Maureen Rouhi, Editor-in-Chief of *Chemical & Engineering News. Chem. Inf. Bull.* 65 (3). http://bulletin.acscinf.org/node/493.

Baykoucheva, S., Twiss-Brooks, A., 2012. Talking about eScience, libraries, and other things: an interview with Andrea Twiss-Brooks. *Chem. Inf. Bull.* 64 (1). http://bulletin.acscinf.org/node/306.

Garfield, E., 1979. *Current Contents*: Its impact on scientific communication. *Interdiscip. Sci. Rev.* 4 (4), 318–23. http://www.garfield.library.upenn.edu/essays/v6p616y1983.pdf.

Garfield, E., 2014. Home Page. Retrieved November 14, 2014, from http://www.garfield.library.upenn.edu/.

Gordin, M.D., 2004. *A Well-ordered Thing: Dmitrii Mendeleev and the Shadow of the Periodic Table*. Basic Books: New York, NY.

Scerri, E.R., 2006. *The Periodic Table: Its Story and Its Significance*. Oxford University Press: New York, NY.

Scientific communication in the digital age

> Modern dialog formats in science communication are reminiscent of a culture of public discourse and involvement in past centuries.
>
> *Könneker and Lugger (2013)*

> Nobody reads journals. People read papers.
>
> *Vitek Tracz (Rabesandratana, 2013)*

> Peer review at its best is a continual process of critique and assessment.
>
> *Marincola (2013)*

2.1 Introduction

The advances in digital technologies, the rapid growth in the numbers of submissions of papers to scientific journals, problems with peer review, emergence of social media as a vehicle of communication in science, and alternative metrics for evaluating the impact of research are causing a dramatic change in scientific publishing (Harley, 2013; Könneker and Lugger, 2013; Rabesandratana, 2013).

The development of new information architecture and Semantic Web technologies that are based on open access, open data, and open standards allow users to exchange content online and collaborate with people of similar interests. The web makes it possible to publish, share, and link text, data, images, video, and other artifacts that can be aggregated and stored in a cloud. In this new inter-connected digital environment, known as Scholarship 2.0, scientists will be able to use tools provided by the new Internet technologies to collaborate and build new knowledge on the existing science.

The next sections of this chapter discuss how the new technologies and ideas are challenging the traditional forms of scientific publishing, changing the whole STEM publishing field.

2.2 Challenging the traditional scientific publishing model

It is obvious to many that the current publication system has become in many respects dysfunctional, which Priem and Hemminger attributed to "the tight coupling of the journal system: the system's essential functions of archiving, registration, dissemination, and certification are bundled together and soloed into … individual journals. This tight coupling makes it difficult to change any one aspect of the system, choking out innovation" (Priem and Hemminger, 2012). They proposed adopting a "decoupled journal (DcJ)" system, in which the functions are "unbundled" and scholars can use many different services to deposit their articles and have them indexed and reviewed by different agencies, taking advantage of new web technologies, tools, and networks.

The peer review as practiced by traditional journals has been widely criticized, with alternative models emerging to challenge it (Bornmann and Daniel, 2009). Open-access pioneer Vitek Tracz believes that anonymous peer review is "sick and collapsing under its own weight" (Rabesandratana, 2013). He has launched an open-access journal, *F1000Research*, in which articles and all supporting data can be reviewed by eligible peers as soon as they are posted online.

In an editorial (*Are we refereeing ourselves to death? The Peer-Review System at its limit*), François Diederich, chairman of the editorial board of one of the most respected chemistry journals, *Angewandte Chemie*, described the situation with peer review in this way:

> … Since all manuscripts need to be reviewed, the requests for referee reports become increasingly frequent. It becomes impossible to serve all these requests as the scientists also need to do research and teaching and fulfill other duties, depending on their employment at a university, non-university institutions, or in industry …

He pointed out to the unsustainability of peer review at a time when research is taking place on an "unprecedented scale" and "the rapid rise of China in the last two decades has contributed to an enormous growth in the number of publications" (Diederich, 2014). This is how Maureen Rouhi, former editor of the ACS journal *Chemical & Engineering News* (Rouhi, 2014), described the contributions of countries in Asia to the global research and publishing enterprise:

> R&D in Asia is growing at a rapid rate. At the American Chemical Society, evidence of that comes from the publishing services: Submissions to ACS journals from India, South Korea, and China grew at annual compounded rates of 17.3%, 16.6%, and 14.7%, respectively, compared with 5.4% from the U.S., during 2008–12. Researchers in Asia are significant users of ACS information services, including SciFinder. The databases that underpin SciFinder increasingly are based on molecules discovered in Asia. China now leads the world in patent filings.

Reviewing papers for publication has become a significant burden for many scientists, who have many other responsibilities—supervising graduate students, attending seminars and meetings, working on committees, reviewing grants, etc. Several thousand scientists who were surveyed by the National Science Board listed the following top bureaucratic burdens as predominant: proposals, progress reports, agency-specific requirements, effort reporting, data sharing, finances, conflict of interest reporting, human and animal subject protections, and biosafety (Widener, 2014). It might not look so improbable, then, for an article sent for review to a respected but busy researcher to wind up on the desk of a postdoc or even a graduate student.

Another problem with peer review is that it is often difficult to find scientists who are experts in a very specific and narrow field. Then, the reviewers that are selected (and these could be very well respected scientists) may not be working in the particular area of the research reported in the paper. Such reviewers have to rely on their general knowledge of the field rather than on any direct experience, thus making the evaluation less accurate and of a general type.

The publishing field in which scientific knowledge is filtered by peer review and other journal selection processes has moved toward a more open environment (Delgado López-Cózar et al., 2013), and it is important that the publishing system and the academic institutions create incentives for reviewers and reward them for their work, because reviewing reflects well on the department and the organization as a whole.

2.3 The impact of the Open Access Movement on STEM publishing

The Budapest Open Access Initiative of February 14, 2002 (Budapest Open Access Initiative, 2014), defined open access (OA) as "… free availability on the public Internet, permitting users to read, download, copy, distribute, print, search or link to the full text of these articles, crawl them for indexing, pass them as data to software or use them for any other lawful purpose …"

The OA movement has triggered much broader changes in society, in general, and in science, in particular, than expected (Lagoze et al., 2012; Suber, 2012, 2014). Its benefits are now generally accepted, as more people have access to research findings and researchers get more citations of their articles. Increased public interest in science information is quickly and dramatically changing the scientific discourse. Content previously available only to specialists can now be accessed by a wider audience. Patients now often go to their doctors carrying copies of articles they have downloaded from scientific journals.

The culture in a discipline and how researchers in a particular field perform their research play a significant role in choosing outlets for disseminating their work. Faculty still depend for promotion on publishing in high-impact journals that attract the majority of the better papers (Harley, 2013). In some disciplines, researchers have cultivated strong relationships with their professional organizations and are more likely to submit papers to their journals. This is particularly true for chemists, for whom having an article accepted for publication by a reputable ACS journal would be like winning a gold medal.

This is how Bryan Vickery, former editorial director of the OA publisher Chemistry Central, viewed the differences between the disciplines (Baykoucheva, 2007):

> The open access movement has grown quickly over the past few years and took off first in the biomedical sciences. Here, open resources like PubMed and GenBank allowed biomedical researchers to understand the benefits of open access. Other fields have not had that advantage and most of the important resources for chemists are still locked behind subscription barriers. Chemists have traditionally been conservative (compared to biomedical researchers and physicists) in challenging the status quo, but

with some areas of chemistry on the decline, and boundaries between disciplines blurring rapidly it is time to act. There has been increasing recognition that the benefits of open access for the publication of original research apply in all fields, and certainly foster collaborations in multidisciplinary areas.

Together with the benefits of OA, some new trends have emerged that are detrimental to its reputation. The number of predatory OA journals is increasing, as judged by the list of such journals compiled by Jeffrey Beall (Beall, 2014).

2.4 New models of scientific communication and publishing

As a result of the OA movement and the development of new web technologies, the publishing field has become more diverse and open. Technology is changing how users search for and view content. The advances in technology created the possibility of developing new authoring tools that allow anyone to become a publisher—in a matter of hours. Many universities, academic libraries, and individual scholars are taking advantage of these tools so that research findings are published faster and independently of major publishers. Some of the new models of publishing and the organizations involved in developing and supporting such models are discussed below:

arXiv (arxiv.org) provides free e-prints in physics, mathematics, computer science, quantitative biology, quantitative finance, and statistics.

The Association of Research Libraries (ARL) (www.arl.org) is a nonprofit organization that advocates OA. It includes research libraries in the United States and Canada and is one of the original founders of SHARE (Shared Access Research Ecosystem), discussed below.

bepress (Berkeley Electronic Press) (www.bepress.com) offers communication and publishing services for academic institutions. It has created Digital Commons, the leading institutional repository (IR) platform for universities, colleges, law schools, and research centers.

bioRxiv (biorxiv.org) archives and distributes unpublished preprints in the life sciences. Once an article is posted, it can be cited and cannot be removed.

BioMed Central (www.biomedcentral.com) is a publisher of OA peer-reviewed journals in science, technology, and medicine, making articles freely available upon publication.

CHORUS (Clearinghouse for the Open Research of the United States) (www.chorusaccess.org) provides tools and resources for discovery and preservation of research data and supports initiatives for increasing public access to peer-reviewed publications with articles based on federally funded research.

CrossRef (www.crossref.org) is an association of scholarly publishers that maintains a citation-linking network covering millions of journal articles and other content items such as book chapters, data, theses, and technical reports. It is involved in developing new tools and infrastructure to support scholarly communications.

Digital Science (www.digital-science.com) is a company developing innovative technology whose goal is to make scientific research more efficient. It is a major player in creating tools for bibliographic management (ReadCube), Altmetrics (Altmetric), repositories (figshare), laboratory project management (Projects and Labguru), data management (Symplectic Elements), science funding resources (Dimensions for Funders and Dimensions for Universities), and collaborative publishing systems (Overleaf).

eLife Lens (lens.elifesciences.org) uses a technology that takes into account how users search for content and view it. Since readers rarely read an article from the beginning to the end—they usually take a quick look at the paper and then focus on particular parts that are of interest to them—eLife Lens displays the different elements and sections of an article (e.g., text, figures, tables, supplemental data, and references) in individual panels, which allows users to view them simultaneously.

F1000Research (f1000research.com) is an OA journal that provides immediate publication of articles and makes them available for open peer review. It uses invited reviewers, but this happens after the article is made available to the public online, and anybody can review and comment on the presented results and raw data.

figshare (figshare.com) is a repository for many different kinds of files (e.g. figures, datasets, media, papers, posters, and presentations), which can be uploaded and visualized in a browser.

In **FundRef** (www.crossref.org/fundref), publishers deposit funding information from scholarly articles and other contents. FundRef Registry uses a taxonomy of international funders' names provided by Elsevier. The funding data are publicly available through CrossRef.

Horizon 2020/EU Framework Program for Research and Innovation (ec. europa.eu/programmes/horizon2020) is the biggest research and innovation program of the European Union, whose goal is to support research and innovation.

IPython Notebook (ipython.org/notebook.html) is an example of the new powerful "direct-to-publish" authoring tools that are changing users' experiences with scientific information. It allows the user to combine code execution, text, mathematics, plots, and media into a single document using interactive Web-based tools.

The Library Publishing Coalition (www.librarypublishing.org) is a library-led project aimed at supporting publishing by libraries.

National Institutes of Health/National Center for Biotechnology Information (www.ncbi.nlm.nih.gov). The rapid developments in molecular biology required different archiving and publishing models, which led to the creation of the web-based genomic and proteomic databases (GPD), GenBank and Protein Data Bank (Brown, 2003). GPD is used for storage and retrieval, as well as for depositing vast volumes of molecular biology information. GPD is free and the information deposited in it is not subject to peer review.

The Office of Science and Technology Policy (OSTP) (www.whitehouse.gov/administration/eop/ostp) advises the president of the United States on science and technology issues.

Open Access Scholarly Publishers Association (OASPA) (oaspa.org) develops new publishing models that support OA journal and book publishing. Its mission is also to educate researchers and the public on the benefits of OA publishing.

OpenAIRE (Open Access Infrastructure for Research in Europe) (www.openaire.eu) is an initiative of the European Union supporting a large-scale shared archive network of aggregated digital repositories for datasets and other kinds of scientific outputs from many disciplines.

ORCID (orcid.org) provides a unique identifier that disambiguates authors' names. It is discussed in more detail in Chapter 15 of this book.

PeerJ (peerj.com) is an OA peer-reviewed journal providing low-cost publishing plans to authors. It publishes articles in the biological and medical sciences and allows readers to ask and answer questions, as well as make comments and annotations about them. It also maintains a preprint server, **PeerJ PrePrints**.

Peerage of Science (www.peerageofscience.org) is a new model for peer review in publishing, which uses qualified reviewers to evaluate manuscripts before they are submitted to any journal. It also allows the writing of reviews of the peer reviews. Journals that subscribe to this service can accept already reviewed articles, and the authors also have the option of sending them to other journals.

PLoS Labs Open (www.ploslabs.org) is building new tools for disseminating and evaluating research.

Publons (publons.com) is a platform for reviewers to post their reviews and get credit for them. This social networking website provides a platform for sharing and discussing peer review of academic publications.

PubMed Central (www.ncbi.nlm.nih.gov/pmc) is a digital repository that archives full-text scholarly articles in the biomedical and life sciences journal literature and makes them freely available. It is linked to other NCBI (National Center for Biotechnology Information) databases. The NIH public access policy requires that research papers reporting research funded by the National Institutes of Health must be available to the public for free through PubMed Central within 12 months of publication (publicaccess.nih.gov).

Public Library of Science (PLoS) (www.plos.org) is an OA nonprofit scientific publisher. It has proposed the model of article-level metrics (ALMs) (SPARC, 2014). PLoS publishes OA journals and blogs in biology and medicine and provides resources and tools to facilitate the understanding and application of ALMs.

The Public Knowledge Project (PKP) (pkp.sfu.ca) is a multiuniversity initiative for developing (free) open-source software to improve the quality of scholarly publishing. The Open Journal Systems (OJS) is PKP's journal management and publishing system that is freely available to journals. Based on open-source software installed and controlled locally, it is aimed at expanding and improving access to research.

Research Data Alliance (RDA) (rd-alliance.org) is an international organization that connects researchers and enables open sharing of data across technologies, disciplines, and countries.

ROARMAP (Registry of Open Access Repositories Mandatory Archiving Policies) (roarmap.eprints.org) provides a list of open-access mandates.

Rubriq (www.rubriq.com) is a new form of peer review, which prescreens papers that are submitted for publication before they enter the publication process. It is used by some publishers as the only peer-review evaluation.

SHARE (SHared Access Research Ecosystem) (Association of Research Libraries, 2014) is an OA research initiative proposed by the ARL, the Association of American Universities (AAU), and the Association of Public and Land-grant Universities (APLU). Its network of cross-institutional digital repositories will accept papers and associate them with datasets resulting from research funded by federal agencies.

SHERPA/RoMEO (www.sherpa.ac.uk/romeo) and **SHERPA/JULIET** (www.sherpa.ac.uk/juliet) are organizations that maintain databases with publisher copyright and research funders' policies, respectively.

ScienceOpen (www.scienceopen.com) calls itself a "research and publishing network." Started by scientists in 2013 in Leipzig, Germany, it is an independent company with offices in Berlin and Boston that is committed to OA.

SPARC (www.sparc.arl.org) is a nonprofit international organization dedicated to promoting ALMs (SPARC, 2014). The mission of SPARC is to "expand the distribution of the results of research and scholarship in a way that leverages digital networked technology, reduces financial pressures on libraries, and creates a more open system of scholarly communication" (Joseph, 2014).

2.5   Use of social media in scientific communication

Researchers are increasingly using many new venues to communicate with their peers (Fausto et al., 2012; Osterrieder, 2013). Building a social-media presence has become a necessity for making researchers' work more visible. Blogs, social bookmarks, free reference managers, Facebook, Instagram, Twitter, and other social sites are making scientific research much more visible and transparent (Brossard and Scheufele, 2013; Fausto et al., 2012; Shuai et al., 2012).

Mewburn and Thomson performed content analysis of 100 academic blogs and identified nine main types of academic activity on the blogs (Mewburn and Thomson, 2013): self-help (e.g. advice to graduate students), descriptions of academic practices, technical advice, academic culture critique, research dissemination (e.g. summaries of research results presented in plain language), career advice, personal reflections, information (e.g. advertising conferences, notifications of new papers, reports, or seminars), and technical advice.

Whether social media will be used by researchers depends on the discipline (Haustein et al., 2014). An investigation of how researchers from 10 disciplines used Twitter showed that biochemists retweeted significantly more often than researchers from the other disciplines (Holmberg and Thelwall, 2014). The article showed that Twitter was used for scholarly information most often by those working in the fields of biochemistry, astrophysics, cheminformatics, and digital humanities.

Another study analyzed the use of blogs on the Research Blogging (RB) platform to see how researchers from different disciplines are contributing to the blogosphere (Fausto et al., 2012). The article showed that the biologists were the most active bloggers (36%), followed by health professionals (15%) and psychologists (13%). Besides these differences by discipline, there could be other reasons people do not use social media, such as "fear of appearing unprofessional, posting something wrong or being misunderstood, or lack of confidence in your computer skills" (Osterrieder, 2013).

An Ithaka S + R project (Long and Schonfeld, 2014) looked at the research habits and needs of academic chemists and found that "despite their heavy use of technology for research, many academic chemists have been slow to adopt new models of sharing data and research results such as online repositories and open access publishing." Using social networks and posting works in institutional repositories make researchers more visible. This also allows them to find people with similar interests and to expand their professional networks (Brossard and Scheufele, 2013).

2.6 Conclusion

For many years, the supporters of OA had to overcome prejudice and even outright opposition from some commercial and professional society publishers. After resisting it for years, many of the for-profit publishers now have OA journals and are providing new alternative publishing solutions, such as Gold OA (free and open dissemination of original scholarship) and Green OA (in which free and open dissemination is achieved by archiving and making freely available copies of scholarly publications that may or may not have been published previously).

Researchers have become more receptive to the idea of publishing their results in OA journals, with the whole STEM publishing field transitioning more fully to it. Funding agencies and academic institutions are increasingly pushing scientists to publish in OA journals and make research findings available to everyone. They encourage researchers to reach out to the general public and write for outlets (blogs and web pages) other than traditional scientific journals. The funding agencies in the United Kingdom, for example, are now requiring grant applicants to write their proposals in such a way that the general public can understand the impact of their research (Mewburn and Thomson, 2013).

Scientists want to have their works to be disseminated as broadly as possible, but as a study has shown, they are not even posting their publications in institutional repositories, which can quickly make them available to everyone (Davis and Connolly, 2007). Creating profiles in Google Scholar, Academia.edu, and ResearchGate.net, or uploading PowerPoint presentations on SlideShare or videos on YouTube and Vimeo can significantly increase the visibility of an author. These sites show statistics such as "views," "downloads," and "likes" that provide information about the attention the posted works have attracted. Registering for the unique author identifiers ResearcherID and ORCID (discussed in detail in Chapter 15 of this book) would be beneficial to all authors, especially those whose names are transliterated from other languages or have appeared in different forms on their publications.

Discontent with current peer-review practices has fueled interest in developing new forms of communication. There are ongoing efforts to detach peer review from the journals. Some new publishing initiatives are experimenting with "free-for-all" commenting systems, but it is not clear, yet, whether people will be willing to participate in these. Even with the new tools and approaches, peer review will never be flawless, but the question is how to improve the methods of evaluation and find better measures of scientific quality so as to be able to screen papers both before and after publication (McNutt, 2013). The developments discussed in this chapter have facilitated the emergence of a diverse information landscape, which not only presents challenges but also creates excitement for researchers, publishers, and the general public, as more and more scientific information becomes available, for the benefit of us all.

References

Association of Research Libraries, 2014. SHared Access Research Ecosystem (SHARE). Retrieved July 31, 2014, from http://www.arl.org/focus-areas/shared-access-research-ecosystem-share#.U9pSa6P_ln8.

Baykoucheva, S., 2007. Paving the road to more open access for chemistry: interview with Bryan Vickery, Editorial Director of Chemistry Central. *Chem. Inf. Bull.* 59 (1), 13–15. http://acscinf.org/content/paving-road-more-open-access-chemistry.

Beall, J., 2014. List of Predatory Publishers 2014. Retrieved July 26, 2014, from http://scholarlyoa.com/2014/01/02/list-of-predatory-publishers-2014.

Bornmann, L., Daniel, H.D., 2009. Reviewer and editor biases in journal peer review: an investigation of manuscript refereeing at *Angewandte Chemie International Edition. Res. Eval.* 18 (4), 262–72. http://rev.oxfordjournals.org/content/18/4/262.

Brossard, D., Scheufele, D.A., 2013. Science, new media, and the public. *Science* 339 (6115), 40–1. http://rev.oxfordjournals.org/content/18/4/262.

Brown, C., 2003. The changing face of scientific discourse: analysis of genomic and proteomic database usage and acceptance. *J. Am. Soc. Inf. Sci. Technol.* 54 (10), 926–38.

Budapest Open Access Initiative, 2014. Budapest Open Access Initiative. Retrieved August 1, 2014, from http://www.budapestopenaccessinitiative.org.

Davis, P.M., Connolly, M.J.L., 2007. Institutional repositories: evaluating the reasons for non-use of Cornell University's installation of DSpace. *D-Lib Magazine* 13 (3/4). http://www.dlib.org/dlib/march07/davis/03davis.html.

Delgado López-Cózar, E., Robinson-García, N., Torres-Salinas, D., 2013. Science communication: flawed citation indexing. *Science* 342 (6163), 1169. http://www.sciencemag.org/content/342/6163/1169.2.

Diederich, F., 2014. Are we refereeing ourselves to death? The peer-review system at its limit. *Angew. Chem. Int. Ed.* 52 (52), 13828–9. http://onlinelibrary.wiley.com/doi/10.1002/anie.201308804/abstract.

Fausto, S., Machado, F.A., Bento, L.F.J., Iamarino, A., Nahas, T.R., Munger, D.S., 2012. Research blogging: indexing and registering the change in science 2.0. *PLoS One* 7 (12), 10. http://journals.plos.org/plosone/article?id=10.1371/journal.pone.0050109.

Harley, D., 2013. Scholarly communication: cultural contexts, evolving models. *Science* 342 (6154), 80–2. http://www.sciencemag.org/content/342/6154/80.

Haustein, S., Peters, I., Sugimoto, C.R., Thelwall, & M., Lariviere, V., 2014. Tweeting biomedicine: an analysis of tweets and citations in the biomedical literature. *J. Assoc. Inf. Sci. Technol.* 65 (4), 656–69. http://onlinelibrary.wiley.com/doi/10.1002/asi.23101/abstract.

Holmberg, K., Thelwall, M., 2014. Disciplinary differences in Twitter scholarly communication. *Scientometrics* 101 (2), 1027–42. http://dx.doi.org/10.1007/s11192-014-1229-3.

Joseph, H., 2014. The Evolution of Open Access: What Might Happen Next? Future of the Research Library Speaker Series. Retrieved August 1, 2014, from http://hdl.handle.net/1903/15042.

Könneker, C., Lugger, B., 2013. Public Science 2.0—back to the future. *Science* 342 (6154), 49–50. http://dx.doi.org/10.1126/science.1245848.

Lagoze, C., Van De Sompel, H., Nelson, M., Warner, S., Sanderson, R., Johnston, P., 2012. A web-based resource model for scholarship 2.0: object reuse & exchange. *Concurr. Comput. Pract. Exp.* 24 (18), 2221–2240. http://dx.doi.org/10.1002/cpe.1594.

Long, M.P., Schonfeld, R.C., 2014. Supporting the Changing Research Practices of Chemists. Ithaka S+R. Retrieved July 31, 2014, from http://www.sr.ithaka.org/research-publications/supporting-changing-research-practices-chemists.

Marincola, E., 2013. Science communication: power of community. *Science* 342 (6163), 1168–9. http://dx.doi.org/10.1126/science.342.6163.1168-b.

McNutt, M., 2013. Improving scientific communication. *Science* 342 (6154), 13. http://dx.doi.org/10.1126/science.1246449.

Mewburn, I., Thomson, P., 2013. Why do academics blog? An analysis of audiences, purposes and challenges. *Stud. High. Educ.* 38 (8), 1105–19. http://dx.doi.org/10.1080/03075079.2013.835624.

Osterrieder, A., 2013. The value and use of social media as communication tool in the plant sciences. *Plant Methods* 9 (1), 26, info:pmid/23845168.

Priem, J., Hemminger, B.M., 2012. Decoupling the scholarly journal. *Front. Comput. Neurosci.* 6, 19. http://dx.doi.org/10.3389/fncom.2012.00019.

Rabesandratana, T., 2013. The seer of science publishing. *Science* 342 (6154), 66–7. http://dx.doi.org/10.1126/science.342.6154.66.

Rouhi, M., 2014. Next stop: Asia. *Chem. Eng. News* 92 (13), 3.

Shuai, X., Pepe, A., Bollen, J., 2012. How the scientific community reacts to newly submitted preprints: article downloads, Twitter mentions, and citations. *PLoS One* 7 (11), 8. http://dx.doi.org/10.1371/journal.pone.0047523.

SPARC, 2014. Article-Level Metrics. Retrieved July 31, 2014, from http://www.sparc.arl.org/initiatives/article-level-metrics.

Suber, P., 2012. *Open Access*. The MIT Press: Cambridge, MA.

Suber, P., 2014. Open Access Overview. Retrieved July 16, 2014, from http://legacy.earlham.edu/~peters/fos/overview.htm.

Widener, A., 2014. Paperwork paralysis. *Chem. Eng. News* 92 (22), 20–21.

Ethics in scientific publishing 3

3.1 Introduction

The first time I heard about the "Kekulé Riddle" was during a lecture that Dr. Alfred Bader, the founder of the Aldrich Chemical Company (now Sigma-Aldrich), gave at the Department of Chemistry and Biochemistry of the University of Maryland, College Park. Dr. Bader presented evidence that when the German chemist Friedrich August Kekulé published his paper on the benzene ring, he had already seen similar models proposed by an Austrian chemist, Josef Loschmidt.

According to the legend, Kekulé came to the idea of this particular (ring) structure of benzene after dreaming of a snake eating its own tail. His theory was published in 1865—four years after Loschmidt had already proposed a similar structure, which he published in a little-known book. Archibald Scott Couper was the first to propose the theory of the tetravalence of carbon, but it was Kekulé who published it just a month earlier (in May of 1858) than Couper. After the lecture, I had the opportunity to interview Dr. Bader for the *Chemical Information Bulletin* (Baykoucheva, 2007) and asked him this question:

> For chemists who have always been taught that Kekulé was the one who first recognized that carbon is tetravalent and also proposed a theory, according to which benzene forms a ring, it is quite shocking to learn from your presentations that, in fact, other people—Archibald Scott Couper and Johann Josef Loschmidt, respectively, were the first to have these ideas, but their contributions remained unknown. Why are you so passionate about this issue—are you on a mission to re-write the chemistry books?

This is what Dr. Bader said:

> I would simply like the truth to be known. There is almost certainty that Couper submitted his manuscript on the tetravalence of carbon before Kekulé did. And there is absolute certainty that Loschmidt illustrated over a hundred circular aromatic structures in his book of 1861, five years before Kekulé's paper. And Kekulé saw this book no later than January 1862 when he wrote about it to Erlenmeyer. Am I on a mission to re-write chemistry books? Well, if wanting the truth to be known is being on a mission, yes.

3.2 Are we ever going to know the truth?

We may never know the truth, though, judging by what Michael Gordin, a historian at Princeton University, who has done a lot of research on pseudoscience, said when I asked him about this and other cases of disputed authorship in chemistry (Baykoucheva, 2011):

> I am not interested in deciding who was "right"—I don't think historians are in the business of awarding prizes or credit—but the fact that this fight took place, and the kinds of arguments Mendeleev and Meyer used to argue for who was first, makes for a fascinating story to uncover. For better or worse, our system of assigning credit in the sciences centers on priority, and the historian is obligated to explore why that particular system emerged, and what its consequences have been. With respect to Couper, Butlerov, Kekulé, and others—I'm afraid I am a spectator in that historiography and am not going to weigh in on one side or the other, but I can tell you my own particular approach to this kind of question. The fact remains that Kekulé was awarded the credit by his peers. I am personally more interested in why they thought he should receive the credit, rather than in adjudicating whether they were correct or incorrect in doing so.

The scientific publishing enterprise relies strongly on the ethics of the scientific community. With the tremendous pressure on researchers to publish and the huge competition to have papers accepted by high-impact journals, though, scientific fraud is on the rise (Crocker and Cooper, 2011; He, 2013; Maisonneuve, 2012; Matias-Guiu and Garcia-Ramos, 2010; Noyori and Richmond, 2013; Roland, 2007; Steen, 2011). More and more new cases of unethical behavior in researchers are coming to light from the media and from Retraction Watch (He, 2013; Retraction Watch, 2014b), a website monitoring the retractions of articles by scientific journals. Fabricating results in medical research is particularly dangerous, as it could affect the health of many people (Steen, 2012; Zeitoun and Rouquette, 2012).

3.3   Biases of editors

The editors and reviewers of scholarly journals try to do their best to identify areas of concern when reviewing the manuscripts submitted for publication, but they are not immune to bias. A study examined the peer-review process at the journal *Angewandte Chemie International Edition* and looked for evidence of potential sources of bias by analyzing the referees' recommendations and the editors' decisions to accept or reject manuscripts submitted for publication (Bornmann and Daniel, 2009). As the results show, "the number of institutions mentioned in the Acknowledgements of a manuscript, the share of authors having institutional affiliations in Germany, the institutional address of the referee (in Germany or not in Germany), and 'author-suggested referee for the manuscript' have statistically significant effects on the referees' recommendations … the number of institutions that are mentioned in the Acknowledgements and the share of authors having institutional affiliations in Germany are potential sources of bias in the editors' decisions."

Editors may have other biases, too. Several years ago, I sent an inquiry to the editor of a major biomedical open access (OA) journal. I was working on an article showing the benefits of using some chemistry databases, such as SciFinder, to retrieve literature on drugs and wanted to know whether they would consider publishing an article on this topic. The response from the editor was quite unusual—he said that they could not publish an article based on research that had been carried out using a proprietary database to search the literature. So why would a journal that had been publishing research performed using expensive scientific equipment not publish an article reporting results obtained with a proprietary database?

A recent article discussed the issue of "vituperative feedback from peer reviewers" (Comer and Schwartz, 2014). The authors argued that "peer referees have a moral obligation not to humiliate the authors whose work they review." The questionable reviewing practices of some editors and reviewers need to be addressed, to preserve the integrity of the review process.

3.4   Manipulating the impact factor of journals

Some editors have tried to manipulate the impact factor of their journal by following a policy of rejecting articles that may not bring many citations or are not falling in the category of "novelty," considered an important quality by the journal (Arnqvist, 2013). It is also a widely known fact that some editors are putting pressure on the authors of submitted manuscripts to cite articles published in their journal. An editorial published in *Nature Neuroscience* gives the following "advice" to editors who want to boost the IF of their journals (Nature Neuroscience, 1998):

> … publish more reviews, which receive higher citations than original research papers; alter subject coverage in favor of fields with high intrinsic citation rates, such as molecular biology; eliminate topics and sections that generate few citations; and publish controversial editorials. The last method works because when the impact factor is calculated, the numerator is the total number of citations to any item in the journal, whereas the denominator is the number of articles only, and editorials and letters are not normally counted.

It is more common among the medical journals to include reviews, comments, and letters, which inflate the IF. When the IFs of some high-impact biomedical journals such as *The Lancet*, *Annals of Internal Medicine*, and *The New England Journal of Medicine* were recalculated (the authors divided the total number of the cited articles by the total number of articles published, without excluding any articles), the IFs of these journals dropped by 30–40% and their ranking in *Journal Citation Report* (*JCR*) also went down (Moed and van Leeuwen, 1996).

3.5   Peer-review issues

A news article "Who's afraid of peer review?" published in the journal *Science* incriminated many OA journals for accepting a fraudulent paper—more than half of the 304 journals to which it was sent did that (Bohannon, 2013). The concerns about the quality of some OA journals are legitimate, but the way this experiment was performed (not including, as controls, proprietary journals) has made this "study" look more like a "sting" operation aimed at discrediting the OA journals (Joseph, 2013).

The concerns raised by this article pertain not only to the OA journals but also to other peer-review journals. In spite of the rigorous reviewing practices that highly respected scientific journals like *Science* are using, many of them failed to detect unethical behavior and had to retract many articles that contained fraudulent results (Retraction Watch, 2014b; Steen, 2011).

The case of Jan Hendrik Schön, the young scientist from Bell Labs who in only a few years published 17 papers in prestigious scientific journals, is a good example of how even experts in the field can fail to detect such fraud. The research reported in these papers was considered extraordinary and a breakthrough in physics, as it should have been—the creation of molecular-scale transistors, which was claimed, would be a great achievement. Schön was already talked about as someone on a fast track to the Nobel Prize. As it turned out, the articles were based on fiction, not on actual experiments. A book, "Plastic Fantastic: How the Biggest Fraud in Physics Shook the Scientific World", shows how Schön was able to mislead so many people—the reviewers who did not notice that the same graph had been used repeatedly in several different articles and the editors of *Science*, *Nature*, and other prestigious journals that have published his papers and who had to deal with the fallout from this scandal (Reich, 2009). The fabrication of data by Schön had a devastating effect on many researchers who have wasted years of their careers trying to reproduce his results.

A recent article analyzed one of the biggest fabrications of scientific data, done by two researchers in the field of corrosion science, which resulted in the publication of more than 40 articles in high-impact journals (Khaled, 2013). The author of the article proposed conducting data audits of manuscripts submitted for publication as a way to prevent fraudulent results.

The scientific journals usually require that the authors suggest potential reviewers. Hyung-In Moon, a South Korean plant compound researcher, became famous not for his research but for the way in which he misled the editors of many peer-review journals—he made up reviewers and provided for them e-mail addresses that were actually his own. So the journals, without checking the identities of the "reviewers," sent the articles for review—to their author. The truth came out when one of the editors became suspicious after receiving a review only 24 hours after sending the article—something unheard of in the journal practices. By July 21, 2014, the number of Moon's papers retracted by scientific journals had reached 35 (Retraction Watch, 2014a).

3.6   Detecting scientific fraud

The dramatic increase in the number of ethical violations by authors has forced scholarly publishers to revise their publication processes and incorporate additional measures to prevent such cases before publication. Submitted papers are screened for potential conflict of interest, plagiarism, statistics, and image manipulation. A Yahoo or Gmail address provided by an author for a reviewer is likely to trigger additional investigation.

Scientific misconduct can take many different forms, ranging from inadequate citing to outright fraud, and it is sometimes very subtle and difficult to detect and prove. While the most common area of ethical concern is authorship (Claxton, 2005), there are many other types of unethical behavior, including those listed below (Schanze, 2014):

- *Prior publication*: Presented at a conference or published elsewhere.
- *Plagiarism*: Can take different forms—completely or partially copying text without acknowledging the primary source, borrowing ideas and fragments of text without quotation, and, very often, self-plagiarism.
- *Omitting citations*: Not disclosing that there was another similar work already published.
- *Submitting to multiple journals* at the same time (called "hedging").
- *Data or image fabrication, falsification, or manipulation*: Making up, changing, or omitting data (e.g. manipulating gels).
- *Image integrity and standards*: A certain degree of processing might be acceptable, but not masking the original data (no Photoshop).
- *Authorship ethics*: The author should have made a significant contribution; ghost authorship (e.g. administrators); acknowledge technical staff and data professionals.
- *Conflict of interest*.

Image manipulation is one of the most common "technical" violations observed in submitted papers. Publishers require that any documentation should be faithfully representative of those originally captured and that authors should not move, remove, introduce, obscure, or enhance any parts within an image or even adjust contrast. If the images appear to show discontinuity between features, hard lines, similarity between features, spliced lines, contrast adjustment, specific editing, and other modifications, the paper raises concerns. Have object touching tools been used? Is this one image or is it composed of separate pieces? Papers suspected of such violations are put on hold, until the authors provide additional explanations. If unethical behavior is proved, there will be more serious consequences for the author(s) that may include paper rejection, banning from further submission, and notification of institutions.

It is fair to say, though, that in most cases, images have been manipulated not because the authors were trying to mislead the editors and the readers, but because they wanted to improve the presentation of their results in the paper. As far as plagiarism is concerned, most of the time (about 80%), plagiarism is actually self-plagiarism (I have written it; why not use it?).

A sales representative from a vendor of electronic laboratory notebooks (ELNs) told me recently that the main reason pharmaceutical and other companies wanted to adopt ELNs in their workflow was to prevent scientific fraud. When using an ELN, everything that is recorded as part of an experiment—procedures and data—cannot be changed or erased. Each experiment is certified by a witness, who could be a coworker or a supervisor, and any changes made in the record of an experiment will create a new version, keeping the initial record intact.

Omitting experimental data that do not fit into a preliminary hypothesis or that contradict previous research results that have already been published is another way of fudging with data. How many discoveries, though, have been made just because such outliers had not been ignored? Not citing previously published results that either contradict your data or are exactly the same, which makes publishing your data un necessary, is also unethical.

Sometimes, the presence of wrong data in a publication is not due to fraud, but to artifacts observed in experiments. As a visiting scientist many years ago, I joined a group of researchers who had been studying a macrophage receptor for a cell-wall component of *Escherichia coli*. The putative receptor had been isolated and sent to another lab for further analysis. When the results from this analysis came back, we learned that the sample sent for analysis was actually fetuin, a bovine serum protein that was present in the growth medium in which the macrophages had been cultivated. The group had spent several years working on isolating and studying this "receptor" and had published many papers on it. This is a situation that many scientists may find themselves stuck in—it is not easy to admit to the world that something you had believed in and had spent years on turned out to be an artifact.

As mentioned above, misattributed authorship presents one of the most serious violations. With respect to the alignment of authors in a paper, in some countries, such as the United States, the first author is the one who did the actual experimental work—a graduate student, a postdoc, or any other involved in the work. In some countries, traditionally, the first author is the "boss." Publishers give guidance as to who could be included as an author. But in real life, are these guidelines followed? Disputes among scientists about the arrangement of authors on a paper are quite common. Several years ago, I saw my name on an abstract from a conference presentation, which took me by surprise. I had spent a short period of time working in the lab that produced this conference paper, but nobody had told me that I was included as an author. And, to be honest, I had worked on another project in that lab, not the one reported in this abstract. Evidently, this is not rare, if editors of scientific journals have adopted the practice of sending e-mails to all authors listed on a submitted manuscript.

3.7 How do researchers decide what to cite in their publications?

In an interview with Eugene Garfield (Baykoucheva, 2006), I asked him whether there is a nationality bias in selecting certain articles over others to cite, and this is what he said:

> In certain subjects it is inevitable that national journals will be cited instead of international ones. There used to be significant language barriers, but now that most people write in English I don't think this is a significant factor. I can access articles in English whether they are published in an American or a German journal equally well. Of course, we don't cite Chinese language journals much because we may only know the title of the article in English… There are lots of anecdotal statements, but little systematic data. However, in the past certain medical literature studies claimed that American physicians did not cite their British or European counterparts. I doubt however that you find any serious US chemist ignoring the chemical literature of France or Germany when it is relevant.

3.8   Why do researchers resort to unethical behavior?

Michael Gordin gave a historical perspective on the topic, when differentiating pseudoscience from outright fraud (Baykoucheva, 2011):

> On the one hand, we have the category of 'pseudoscience,' which can be roughly defined as something that is not science but tries very hard to look like science and adopt its methods and approaches. That is not quite the same thing as 'fraud,' which connotes a level of insincerity that one doesn't find, for example, among seventeenth-century alchemists. (There is a third category, the hoax, which is something else again.) Now, as to what can be done about any of these things, I do not have any particular insights. Wherever you find science, you will find something that scientists label pseudoscience; the two always come together. Fraud, if one subscribes to a particular model of psychology, is a matter of incentives, and it is possible that with intensified safeguards, one can reduce its occurrence. But we almost certainly can't eliminate it altogether. Peer review, as you mention, is often put forward as a solution to this problem, and it is likely better than having no safeguard at all—at least this guarantees that a few scientists read over the piece before it is published—but the evidence of recent years has shown that it is far from foolproof in catching fraud. But, as in the case of Schön, eventually the misdeeds come to light. Time seems to be our best tool in this matter.

Whether an article will be accepted for publication sometimes depends on who is reviewing it. Authors can suggest possible reviewers and even give the names of researchers to whom they do not want their article to be sent. This is a fairly recent practice, though, because in the past, authors could not reject potential reviewers. Many years ago, a scientist whose lab was involved in a fierce competition with another group told me that a paper he had submitted to a journal was sent for review to the principal investigator of the competitor group. Being also one of the editors of this journal, the reviewer used his influence to delay the publication of the article, while his group worked hard to submit a paper. As a result, the two papers were published in the same issue of the journal.

Authors have to sign a conflict of interest (or rather, lack of it) clause when submitting a manuscript for publication, but there are cases when relationships between academic researchers and pharmaceutical and other companies are more difficult to prove. When examining the affiliations of authors who have published articles on atorvastatin (also known as Lipitor), I found that a significant number of the top 20 most prolific authors had relationships, either directly (through financial support or employment) or indirectly (through coauthorships), with Pfizer (the company that makes atorvastatin) or with other pharmaceutical companies making similar drugs (Baykoucheva, 2008).

3.9   Organizations involved in preventing unethical behavior

Publishers, funding agencies, and research institutions have become very concerned about scientific misconduct and expanded their efforts to educate researchers, students, editors, and reviewers about accepted ethical rules.

The Office of Research Integrity (ORI) at the U.S. Department of Health and Human Services provides a number of tools that allow the detection of improperly re-used text. ORI's image analysis tools, such as Forensic Droplets, a desktop application in Adobe Photoshop, can be automated to examine the questioned images and groups of images in biomedical science in batch.

The Committee on Publication Ethics (COPE) addresses the concerns of editors of scientific journals about scientific fraud, offers advice on different cases, and has a mission to educate authors and reviewers. COPE maintains Retraction Watch and a database with cases of misconduct.

The Council of Science Editors (CSE) raises awareness among editors and educates authors about unethical behavior.

The Online Resource Center for Ethics Education in Science and Engineering (ORCEESE) of the National Science Foundation provides resources for researchers for the preparation of grant proposals (US National Science Foundation (NSF), 2014).

The National Institutes of Health (NIH) has posted recommendations on conflict-of-interest issues for biomedical journals (Cooper et al., 2006).

CrossCheck is a service provided by **CrossRef** designed to detect plagiarism and help publishers verify the originality of content submitted to them for publication. It compares the text of manuscripts to a unique database of content, which contains the full text of already published articles, books, and conference proceedings from hundreds of publishers.

Some scientific journals now require raw experimental data to be included in the papers submitted for publication (Enserink, 2012; Marusic, 2010). Although fraud cannot be fully avoided by any control system, everything possible should be done to prevent it, because the intentional misconduct of a single author can seriously damage the reputation of a department, an institution, and a publication.

3.10 Conclusion

After spending years and failing to confirm results published in articles, many researchers are finding out that it was not their fault that they could not reproduce the results reported in these papers—because the data have been fabricated. Unethical behavior could have detrimental consequences on researchers' careers, the reputation of scientists, and the use of scientific information.

Plagiarism is a gray area, and there are no common rules and interpretations that cover all cases. Sometimes, it is unintentional—people take notes when reading other publications and forget to distinguish them from the original text. Plagiarism involves a conscientious effort to mislead readers by expropriating the contribution of others. Citing a reference is not enough and other ways should also be used to indicate that the text is a quotation—quotation marks or a substitute for them (e.g., a change in indentation/spacing to indicate a block of quote). To be on the safe side, it's better to just paraphrase and avoid being accused of plagiarism.

Graduate students and postdocs depend on their supervisors for current financial support and recommendations for future jobs. Presenting results that differ from those previously published by the lab or that do not confirm a preliminary hypothesis might turn out to be very detrimental for their careers. Such situations could potentially lead to data manipulation. It is very important that students, from the very beginning of their research, be educated about ethical standards in research and the potential risks of not abiding by these standards.

References

Arnqvist, G., 2013. Editorial rejects? Novelty, schnovelty! *Trends Ecol. Evol.* 28 (8), 448–9. http://dx.doi.org/10.1016/j.tree.2013.05.007.

Baykoucheva, S., 2006. Interview with Eugene Garfield. *Chem. Inf. Bull.* 58 (2), 7–9. http://acscinf.org/content/interview-eugene-garfield.

Baykoucheva, S., 2007. Chemistry and art: the incredible life story of Dr. Alfred Bader. *Chem. Inf. Bull.* 59 (1), 10–12. http://acscinf.org/content/chemistry-and-art-incredible-life-story-dr-alfred-bader.

Baykoucheva, S., 2008. Analyzing the literature on drugs with Web of Science and HistCite: institutional affiliations of the most prolific authors publishing on atorvastatin (Lipitor). Paper presented at the 4th International Conference on Webometrics, Informetrics and Scientometrics (WIS), Berlin, Germany. http://www.collnet.de/Berlin-2008, Can be accessed at http://hdl.handle.net/1903/13940.

Baykoucheva, S., 2011. Political, cultural, and technological impacts on chemistry. An interview with Michael Gordin, Director of Graduate Studies of the Program in the History of Science, Princeton University. *Chem. Inf. Bull.* 63 (1), 50–56. http://www.acscinf.org/content/political-cultural-and-technological-impacts-chemistry.

Bohannon, J., 2013. Who's afraid of peer review? *Science* 342 (6154), 60–65. http://www.umass.edu/preferen/You%20Must%20Read%20This/BohannonScience2013.pdf.

Bornmann, L., Daniel, H.-D., 2009. Reviewer and editor biases in journal peer review: an investigation of manuscript refereeing at *Angewandte Chemie International Edition*. *Res. Eval.* 18 (4), 262–72. http://dx.doi.org/10.3152/095820209x477520.

Claxton, L.D., 2005. Scientific authorship: Part 1. A window into scientific fraud? *Mutat. Res.* 589 (1), 17–30. http://dx.doi.org/10.1016/j.mrrev.2004.07.003.

Comer, D.R., Schwartz, M., 2014. The problem of humiliation in peer review. *Ethic. Educ.* 9 (2), 141–56. http://dx.doi.org/10.1080/17449642.2014.913341.

Cooper, R.J., Gupta, M., Wilkes, M.S., Hoffman, J.R., 2006. Conflict of interest disclosure policies and practices in peer-reviewed biomedical journals. *J. Gen. Intern. Med.* 21 (12), 1248–52. http://dx.doi.org/10.1111/j.1525-1497.2006.00598.x.

Crocker, J., Cooper, M.L., 2011. Addressing scientific fraud. *Science* 334 (6060), 1182.

Enserink, M., 2012. Scientific ethics. Fraud-detection tool could shake up psychology. *Science* 337 (6090), 21–2.

He, T., 2013. Retraction of global scientific publications from 2001 to 2010. *Scientometrics* 96 (2), 555–61. http://dx.doi.org/10.1007/s11192-012-0906-3.

Joseph, H., 2013. *Science* Magazine's Open Access "Sting". Retrieved from http://www.sparc.arl.org/blog/science-magazine-open-access-sting.

Khaled, K.F., 2013. Scientific fraud and the power structure of science. *Res. Chem. Intermed.* 40 (8), 2785–98. http://dx.doi.org/10.1007/s11164-013-1128-x, Epub ahead of print.

Maisonneuve, H., 2012. The management of errors and scientific fraud by biomedical journals: they cannot replace institutions. *Presse Med.* 41 (9 Pt 1), 853–60.

Marusic, A., 2010. Editors as gatekeepers of responsible science. *Biochem. Med.* 20 (3), 282–7. http://dx.doi.org/10.11613/BM.2010.035.

Matias-Guiu, J., García-Ramos, R., 2010. Fraud and misconduct in scientific publications. *Neurologia* 25 (1), 1–4.

Moed, H.F., van Leeuwen, T.N., 1996. Impact factors can mislead. *Nature* 381 (6579), 186.

National Science Foundation (NSF), 2014. Online Resource Center for Ethics Education in Science and Engineering—National Science Foundation (NSF). Retrieved from http://www.nsf.gov/funding/pgm_summ.jsp?pims_id=503490.

Nature Neuroscience, 1998. (Editorial) Citation data: the wrong impact? *Nat. Neurosci.* 1 (8), 641–2. http://stat.smmu.edu.cn/stone/citation%20data-the%20wrong%20impact.pdf

Noyori, R., Richmond, J.P., 2013. Ethical conduct in chemical research and publishing. *Adv. Synth. Catal.* 355 (1), 3–9. http://dx.doi.org/10.1002/adsc.201201128.

Reich, E.S., 2009. *Plastic Fantastic: How the Biggest Fraud in Physics Shook the Scientific World*. Palgrave Macmillan: New York, NY.

Retraction Watch, 2014a. Posts about Hyung-In Moon on Retraction Watch. Retrieved July 21, 2014, from http://retractionwatch.com/category/by-author/hyung-in-moon.

Retraction Watch, 2014b. Retraction Watch. Retrieved July 20, 2014, from http://retraction-watch.com.

Roland, M.-C., 2007. Publish *and* perish. Hedging and fraud in scientific discourse. *EMBO Rep.* 8 (5), 424–8. http://dx.doi.org/10.1038/sj.embor.7400964.

Schanze, K.S., 2014. Ethics in scientific publication: Observations of an editor and recommended best practices for authors. Paper presented at the 247th ACS National Meeting & Exposition, Dallas, TX, United States, March 16-20, 2014.

Steen, R.G., 2011. Retractions in the scientific literature: is the incidence of research fraud increasing? *J. Med. Ethics* 37 (4), 249–53. http://dx.doi.org/10.1136/jme.2010.040923.

Steen, R.G., 2012. Retractions in the medical literature: how can patients be protected from risk? *J. Med. Ethics* 38 (4), 228–32. http://dx.doi.org/10.1136/medethics-2011-100184.

Zeitoun, J.D., Rouquette, S., 2012. Communication of scientific fraud. *Presse Med.* 41 (9 Pt 1), 872–7.

An editor's view: interview with John Fourkas

4

John Fourkas is the Millard Alexander Professor of Chemistry at the University of Maryland, College Park, and holds appointments in the Department of Chemistry and Biochemistry and the Institute for Physical Science and Technology. He earned a BS and MS in Chemistry from the California Institute of Technology and a PhD in Chemistry from Stanford University. He was an NSF postdoctoral fellow at the University of Texas at Austin and the Massachusetts Institute of Technology. He has been a senior editor of *The Journal of Physical Chemistry* since 2002.

Svetla Baykoucheva: What do the editors look for when they first see a manuscript submitted for publication? What would you check first for a new paper?

John Fourkas: Typically, the first thing that an editor would check is whether the subject matter of the paper is appropriate for the journal. The topic should be within the discipline(s) covered by the journal and may need to fit other criteria as well (such as currency and impact). An editor will also check whether the manuscript is suitable to send out for review. Some of the criteria that might be checked include being written in clear English, having high-quality figures, and having an adequate number of references.

SB: How are reviewers selected and vetted for each manuscript?

JF: The selection process for reviewers varies from journal to journal and even editor to editor. I usually select one or more reviewers suggested by the authors and one or more of my own. It is generally helpful if the authors suggest at least five reviewers and list brief qualifications for each. Other reviewers may be chosen based on the editor's knowledge of the subject area, references in the paper, and/or literature/database searches.

Sometimes, after a manuscript has received unfavorable reviews, I receive a complaint from the corresponding author that I have chosen reviewers who are not knowledgeable in the subject area. In my experience, the vast majority of such negative

Managing Scientific Information and Research Data

reviews actually come from reviewers who were suggested by the authors. It therefore is generally not helpful to say this to an editor unless the reviewer has used an unprofessional approach in the review. My experience as an editor (and an author) is that if the reviewer has misunderstood a paper, that is generally a sign that the paper was not written in as clear a manner as might have been imagined by the authors.

SB: What are the most common types of unethical behavior? What are the consequences for authors if they have engaged in such activity?

JF: The answer to this question may depend on the scientific discipline in question, but the type of unethical behavior that most often comes to my attention is plagiarism. For instance, in manuscripts written by authors who are not native speakers of English, I sometimes find sentences in abstracts, introductions, and conclusions that are clearly in a different style from the majority of the manuscript. A Google search often reveals that these sentences were taken from other papers in the literature.

For a number of years, I taught a course on professional skills for new graduate students in chemistry, and one topic that we covered was plagiarism. I gave an assignment in which students were asked to find papers in a hot area of chemistry that was at least five years old. The goal was to try to find instances of plagiarism by looking for such sentences. I asked them not to spend more than 30 minutes on the assignment, and more than 80% were successful in finding examples.

In my experience, this type of plagiarism often arises when students who are not confident of their English abilities are also not well-versed in publication ethics. It is important that lead investigators teach their students about publication ethics and read all manuscripts carefully before submission. This type of infraction may cause a paper to be rejected, and, depending on the severity, it may also result in some sort of sanctions for the authors.

I have also seen many cases in which authors reuse words and/or figures from previous publications. Many authors are not aware that such self-plagiarism is not ethical. The publisher of an article generally holds the copyright, and repeating language or copying figures without attribution is a violation of that copyright. Often, an editor will give an author a chance to rewrite a self-plagiarized paper, depending on the degree of overlap with previous manuscripts.

Scientific misconduct, such as falsification of data, usually is caught by reviewers rather than editors. This type of misconduct is rare, but it is dealt with severely when it comes to light.

SB: What is your journal doing to prevent fraudulent results being published? What tools are you using to detect unethical behavior?

JF: Peer reviewers are the major line of defense in discovering fraudulent results before publication. After publication, the audience of the journal may report fraud as well. In some fields, journals may subject figures and data tables to tests designed to detect some types of fraud as well. In the case of plagiarism, ACS journals routinely check papers that have made it through the first stage of review for overlap with other papers in the literature.

SB: Does the rejection rate of the journal say something about the quality of the articles published in it? Do journals provide information about the rejection rates for their journal?

JF: For better or for worse, impact factors have changed the face of scientific publishing. For example, I visited one university that has a rule that to earn a PhD, the sum of the impact factors for the journals for each article that a student has published must exceed a threshold value. In some countries, publication in a highly regarded international journal is required to earn a graduate degree. Impact factors also play a role in promotions and raises for faculty in many universities around the world.

Given the major impetus to publish in journals with the highest possible impact factor, there is unquestionably a link between impact factor and rejection rate. However, it is also important to note that impact factors vary tremendously from field to field. It is obviously not possible to reduce the importance of a journal to a single number, but such quantification has become increasingly important over the past couple of decades. Although essentially every journal publicizes its impact factor, very few journals reveal their rejection rates, probably for fear of scaring away potential authors.

SB: What do you think of the current state of the peer-review process and how do you see the future of peer review? What do you think of post-publication review?

JF: Overall, I would say that peer review is in good shape. No system is perfect, but so long as editors are careful and reviewers take their jobs seriously, peer review works pretty well. There is certainly a trend toward post-publication review in some fields, but it has not gained wide popularity. This situation might change in the future, but I think that this will depend on the attitudes and interests of the next generations of scientists.

SB: What happens (and does it happen often) when the rejection of a paper is contested?

JF: I suspect that the frequency with which rejections are contested is directly proportional to the impact factor of the journal in question. When the stakes are high, there tend to be fewer alternative venues for publication. From the standpoint of an editor and an author, I have seen the greatest success in contesting rejections when cogent arguments can be made that a reviewer was correct to question some aspect of the work, but that the issues were not as grave as the reviewer believed. This situation usually arises when the author has not been as clear as possible in their writing. One must take on a reasonable tone in any dealing with editors and accept defeat gracefully if need be.

SB: How can researchers increase their chance of getting their papers accepted for publication in good scientific journals?

JF: First, make sure that the topic of your paper is appropriate for the journal to which you submit. Second, write a detailed cover letter that explains to the editor why your paper belongs in that journal. Third, write the best paper that you can and have colleagues review it for you if needed. Finally, recommend qualified reviewers who can give an honest judgement of your work.

Finding and managing scientific information

> Selecting relevant information and suppressing details is the sort of pragmatic fudging everyone does every day. It's a way of coping with too much information. For almost everything you see, hear, taste, smell, or touch, you have the choice between examining details by scrutinizing very closely, and looking at the 'big picture' with its other priorities.
>
> *Lisa Randall (*Warped Passages: Unraveling the Mysteries of the Universe's Hidden Dimensions, *2006)*

5.1 Introduction

Information comes from many different sources and in many different formats. In the past, there was no single search that would take care of everything you could find. As science is becoming more and more interdisciplinary and data-heavy, integration of resources and tools on a single platform is becoming a norm. Such integration allows searching for information across two or more databases. Literature databases are sharing a platform with chemical property and reaction databases, patent sources, and book catalogs. Vendors are competing to offer more and more sophisticated tools to make searches more efficient and allow users to search multiple resources from one point of access (Oprea and Tropsha, 2006; Schenck and Zapiecki, 2014).

5.2 Discovery tools

Libraries and database publishers are trying to introduce discovery tools and features that provide an experience similar to the one users have when searching Google (Parry, 2014). Discovery tools are electronic systems used to bind together different information sources into a single system that can be searched. Results of a search are organized by relevance, and publishers use different priorities for ranking retrieved documents. Ranking often depends on where the search terms match words in the documents. Exact matches have priority over partial matches, and the more often a word appears in a document (especially if the term appears in the title or in the abstract) the higher priority it will be given. Items whose terms match words in the subject headings of an article or book will usually get a higher ranking than those that match words located in the rest of the text.

5.3 "Smart" tools for managing scientific information

Publishers provide different filters or "facets" to refine information. A facet is a specific aspect or description of the item that is somehow associated with the search terms. Such filters allow users to narrow down long lists of references to a more manageable number of results. Literature databases use different categories to group the features that users can apply to filter their results. The categories could be document types (e.g. research articles, reviews, conference papers, books, patents, and dissertations), research/subject areas, authors, group authors, editors, source titles (e.g. journal titles), book series titles, organization/affiliation (e.g. of authors), publication years, funding agencies, languages, and countries/territories.

5.4 Information resources and filtering of information

Some of the most widely used resources for finding scientific literature are Google Scholar, Engineering Village (Compendex), EBSCO, ProQuest, PubMed, SciFinder, ScienceDirect, Scopus, and Web of Science (WoS). Only two of these databases—PubMed and Google Scholar—are free, while the others require paid subscriptions.

Databases that provide information about properties of chemical compounds include ChemSpider, CRC Handbook of Chemistry and Physics, PubChem, Reaxys, SciFinder, and The Merck Index. ChemSpider (from the Royal Society of Chemistry) and PubChem (from NCBI) are free resources, while the others require subscriptions. Sometimes, literature databases reside on the same platform as property resources and scientific tools, which allows the retrieval of a wide variety of information from a single access point (Baykoucheva, 2007, 2011).

5.4.1 Resources from the U.S. National Library of Medicine (NLM)

www.nlm.nih.gov

The National Library of Medicine (NLM) at the U.S. National Institutes of Health (NIH) is the largest biomedical library in the world, which maintains an enormous print collection and produces many electronic information resources. NIH has been at the forefront of the open access movement, maintaining PubMed Central (www.ncbi.nlm.nih.gov/pmc), which makes research funded by federal grants openly available one year after it is published.

5.4.1.1 PubMed/MEDLINE (National Library of Medicine)

www.ncbi.nlm.nih.gov/pubmed

PubMed is a comprehensive free database for biomedical literature that is a major component of the NLM suite of resources (Anders and Evans, 2010). PubMed is based on the MEDLINE database, in which documents are indexed with a controlled vocabulary thesaurus. The "Advanced" search option in PubMed allows the use of "smart" indexing tools to retrieve only articles that are directly related to the topic of the search.

5.4.1.2 PubChem (NCBI)

https://pubchem.ncbi.nlm.nih.gov

PubChem is part of the National Center for Biotechnology Information (NCBI) retrieval system, Entrez. It provides information on the chemical properties and biological activities of small molecules (Baykoucheva, 2007; Sayers et al., 2012). PubChem is organized as three linked databases (PCSubstance, PCCompound, and PCBioAssay), which enable users to find substance information, compound structures, and bioactivity data.

5.4.2 Google Scholar

www.scholar.google.com

Google Scholar (GS) is an interdisciplinary comprehensive resource for finding scientific literature. Its advanced search interface allows users to search for phrases and individual words and limit the search to articles where the search terms were found in the title of the article. Searches can also be narrowed down by author, publication year, journal, and subject categories. If an author's profile is made public, it will appear in GS results when people search for this author's name. GS provides graphic representation of citations to articles over time and calculates several citation metrics that can be updated automatically.

5.4.3 Reaxys (Reed Elsevier Properties SA)

www.reaxys.com

The Reaxys database provides a single platform for searching the world's most extensive collection of organic, organometallic, and inorganic chemistry data. It searches the Beilstein Handbook of Organic Chemistry, Gmelin's Handbook of Inorganic and Organometallic Chemistry, the US Patent and Trademark Office database, and the World and European Patents (Alexander et al., 2014; Buntrock, 2013).

The Beilstein database is the largest database in the field of organic chemistry, covering the scientific literature from 1771 to 1979. It contains experimentally validated information on millions of chemical reactions and substances from original scientific publications. The electronic database was originally based on the Beilstein Handbook of Organic Chemistry, created by the Russian chemist Friedrich Konrad Beilstein in 1881. In 2009, the content of the Beilstein database was incorporated into Reaxys and is maintained and distributed by Elsevier Information Systems in Frankfurt. The Reaxys registered trademark and the database itself are owned and protected by Elsevier Properties SA. In Reaxys, hundreds of fields containing chemical and physical data (such as melting point and refractive index) are available for each substance, accompanied by references to the literature in which the reaction or substance data have been published.

5.4.4 SciFinder (CAS)

www.cas.org

SciFinder is produced by the Chemical Abstracts Service (CAS), a division of the American Chemical Society. SciFinder integrates two large literature databases: the Chemical Abstracts database (CAplus) and the biomedical database MEDLINE (Baykoucheva, 2011; Garritano, 2013). Besides journal articles, it also covers a large variety of other document types such as conference papers, books, dissertations, and patents.

After performing a search, users have to choose whether they want references that contain the search terms exactly the way they were entered or if they prefer to use those in which the terms were used as "concepts." Choosing the option of the references with the terms exactly as entered is appropriate only if the order of words has to be the way it was typed (e.g. "alkaline phosphatase"). In all other cases, the appropriate choice would be 'concepts'. A concept goes beyond the actual meaning of specific words.

SciFinder is an indexed database that searches by default both CAplus and MEDLINE at the same time. The overlap of source titles covered by the two databases creates duplicates, which need to be removed from the search results. The retrieved documents can be analyzed by many categories, such as author name, company organization, and database (choose either CAplus or MEDLINE), document type, index term, CA concept heading, journal name, language, publication year, and supplementary terms.

When selecting keywords, it is not always possible to find the most suitable terms. By analyzing the results by index term, users can adjust their search strategy by including appropriate index terms. For example, when a search is performed using "cancer" as a keyword and the results are analyzed by index term, users will find out that SciFinder uses for "cancer" the index term "neoplasm."

SciFinder also provides access to the CAS Registry File, the largest property information database. For every new chemical compound reported in the literature, CAS creates a number (called CAS Registry Number), which is like a Social Security number for a compound. The name of a chemical compound can be written in many different ways, but the CAS Number is its unique identifier. To search for property information in SciFinder, users can choose to draw a chemical structure or use the molecular formula, the name of a chemical compound, or its CAS Registry Number.

5.4.5   Scopus (Elsevier)

Scopus is the world's largest interdisciplinary indexed database, published by Elsevier. It provides sophisticated tools for analyzing and refining retrieved documents. As with other literature databases, search results in Scopus can be refined by many categories such as publication years, document type (e.g. review), subject area, and other criteria.

Scopus now displays "Mendeley readership statistics" for any article that was downloaded to the bibliographic management program Mendeley, which is now owned by Elsevier. The statistical information includes demographics (e.g. countries), disciplines, and the career levels of the people who have downloaded the article. Scopus Journal Analyzer, which ranks scientific journals using different criteria, is discussed in detail in Chapter 11.

5.4.6   Web of Science (Thomson Reuters)

Web of Science (WoS) is one of the most widely used databases for scientific literature. It is based on the *Science Citation Index*. Its "genealogy" and origins from the Institute for Scientific Information (ISI) (now Thomson Reuters) are discussed in more detail in three other chapters of this book, Chapters 11–13. A short history of ISI was also published recently (Lawlor, 2014). Other products based on the *Science Citation Index* such as *Journal Citation Reports* and *Essential Science Indicators* are discussed in Chapter 11. Finding and filtering results in WoS is similar to what users do in SciFinder, Scopus, EBSCO, and other scientific databases.

More information about chemistry resources is available from other publications (Baykoucheva, 2007, 2011; Garritano, 2013; McEwen and Buntrock, 2014; Currano and Roth, 2014).

5.5   Comparing resources

Researchers now have many options for finding literature and property information. But which one would serve them best? Most researchers in the life sciences and the biomedical field are not aware that their searches could be much more efficient if they use SciFinder rather than PubMed. The benefit of using PubMed is that it is a comprehensive resource for the life sciences and the biomedical field and that it is free. Drugs are chemical substances and searching in SciFinder both MEDLINE and CAplus provides many advantages, as SciFinder carries more document types (e.g. patents, books, and dissertations) than PubMed.

In a study, I compared the performance of CAplus and MEDLINE in retrieving the literature on drugs (Baykoucheva, 2011). The results showed that searches performed in SciFinder, when CAplus and MEDLINE were searched at the same time, retrieved

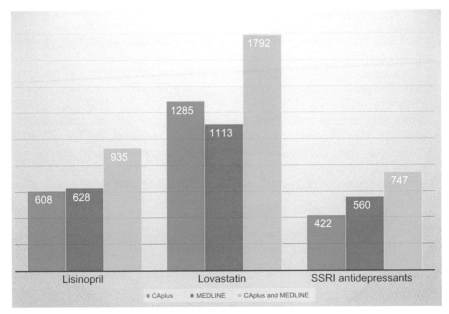

Figure 5.1 Total output of literature on lisinopril, lovastatin, and SSRI antidepressants when CAplus and MEDLINE were searched separately or together (Baykoucheva, 2011).

significantly more documents than when the same searches were performed separately and individually in these databases (Figure 5.1). The latter complemented each other in covering the sources, thus making the literature retrieval much more efficient and comprehensive.

Google Scholar is an interdisciplinary resource with very large coverage of journals. It lacks, though, the more sophisticated features available in the other databases for filtering information.

Scopus and WoS were compared with respect to journal title overlap (Gavel and Iselid, 2008), content and searching capabilities (Salisbury, 2009), and major features (Goodman and Deis, 2007; Jacso, 2005). Two articles analyzed together PubMed and Google Scholar (Anders and Evans, 2010; Shultz, 2007) and another one evaluated the capabilities of PubMed, Scopus, WoS, and Google Scholar (Gomez-Jauregui et al., 2014).

A comparison of the ability of MEDLINE, Scopus, and WoS to retrieve drug literature showed that Scopus performed much better than WoS or MEDLINE in total (Figure 5.2) and annual (Figure 5.3) output (Baykoucheva, 2010). Significant differences were also found in the journal coverage and the total number of documents retrieved from these databases. The article concluded that the best option for retrieving drug literature would be to use both Scopus and WoS, as these two databases complement each other with respect to journal coverage.

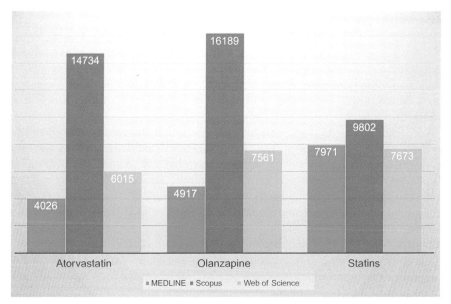

Figure 5.2 Total output of documents published on drugs and retrieved from MEDLINE, Scopus, and Web of Science (Baykoucheva, 2010).

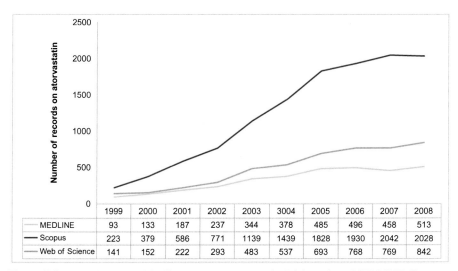

Figure 5.3 Annual output of the literature on atorvastatin (Lipitor) from MEDLINE, Scopus, and Web of Science (Baykoucheva, 2010).

5.6   Conclusion

The integration of scientific databases will continue with more "smart" solutions being provided by database publishers. These new developments will change the role that science librarians play in their institutions, as it will become easier for users to "do it on their own," rather than get training on how to search for information and manage it.

RSS feeds and alerts from publishers provide information about newly published documents on topics of interest. The "Keep Me Posted" service from CAS allows users to find not only regular articles from SciFinder but also "ASAP" papers from both CAplus and MEDLINE, as well as arXiv preprints, theses, ACS meeting abstracts, conference papers, patents, and book chapters. These alerts are much more comprehensive than the RSS feeds and alerts from social networks. If you set up alerts by company affiliation, it retrieves only articles in which the first author has this affiliation. Alerts that will allow the monitoring of the output of individual departments or whole institutions can also be created. Users can also set up alerts and RSS feeds for searches they have performed in Scopus and WoS. A recent article published in *Nature* ("How to tame the flood of literature") mentions some new recommendation services to help researchers keep up with the most important papers and handle information overload (Gibney, 2014).

References

Alexander, J.L., Jürgen, S.-B., Thibault, G., David, E., 2014. The making of Reaxys? Towards unobstructed access to relevant chemistry information. In L.R., McEwan & R.E., Buntrock (Eds.), *ACS Symposium Series: Vol. 1164. The Future of the History of Chemical Information* (pp. 127–48). Washington, DC: American Chemical Society.

Anders, M.E., Evans, D.P., 2010. Comparison of PubMed and Google Scholar literature searches. *Respir. Care* 55 (5), 578–83.

Baykoucheva, S., 2007. A new era in chemical information: PubChem, DiscoveryGate, and Chemistry Central. *Online* 31 (5), 16–20.

Baykoucheva, S., 2010. Selecting a database for drug literature retrieval: a comparison of MEDLINE, Scopus, and Web of Science. *Sci. Tech. Libr.* 29 (4), 276–288. http://dx.doi.org/10.1080/0194262X.2010.522946.

Baykoucheva, S., 2011. Comparison of the contributions of CAPLUS and MEDLINE to the performance of SciFinder in retrieving the drug literature. *Issues in Science & Technology Librarianship* 66 (Summer).

Buntrock, R.E., 2013. Apples and oranges: a chemistry search compares CAS' SciFinder and Elsevier's Reaxys. *Online Searcher* 37 (5).

Currano, J. N., & Roth, D. 2014. *Chemical Information for Chemists: A Primer*. Cambridge, UK: Royal Society of Chemistry.

Garritano, J.R., 2013. Evolution of SciFinder, 2011–2013: new features, new content. *Sci. Techn. Libr.* 32 (4), 346–71. http://dx.doi.org/10.1080/0194262X.2013.833068.

Gavel, Y., Iselid, L., 2008. Web of Science and Scopus: a journal title overlap study. *Online Inf. Rev.* 32 (1), 8–21.

Gibney, E., 2014. How to tame the flood of literature. *Nature* 513 (7516), 129–30. http://dx.doi.org/10.1038/513129a.

Gomez-Jauregui, V., Gomez-Jauregui, C., Manchado, C., Otero, C., 2014. Information management and improvement of citation indices. *Int. J. Inf. Manag.* 34 (2), 257–71.

Goodman, D., 2007. Update on Scopus and Web of Science. *Charlest. Advis.* 8 (3), 15–18.

Jacso, P., 2005. As we may search—comparison of major features of the Web of Science, Scopus, and Google Scholar citation-based and citation-enhanced databases. *Curr. Sci.* 89 (9), 1537–47.

Lawlor, B., 2014. The Institute for Scientific Information: A brief history. In L.R., McEwan & R.E., Buntrock (Eds.), *ACS Symposium Series: Vol. 1164. The Future of the History of Chemical Information* (pp. 109–26). Washington, DC: American Chemical Society.

McEwen, L.R., Buntrock, R.E. (Eds.), 2014. *The Future of the History of Chemical Information*, (ACS Symposium Series, Vol. 1164). Washington, DC: American Chemical Society.

National Center for Biotechnology Information Resource Coordinators, 2013. Database resources of the National Center for Biotechnology Information. *Nucleic Acids Res.* 42 (D1), D7–D17. http://dx.doi.org/10.1093/nar/gkt1146.

Oprea, T.I., Tropsha, A., 2006. Target, chemical and bioactivity databases—integration is key. *Drug Discov. Today* 3 (4), 357–65. http://dx.doi.org/10.1016/j.ddtec.2006.12.003.

Parry, M., 2014. As researchers turn to Google, libraries navigate the messy world of discovery tools. *Chron. High. Educ.* Retrieved April 21, 2014, from http://chronicle.com/article/As-Researchers-Turn-to-Google/146081/.

Salisbury, L., 2009. Web of Science and Scopus: A comparative review of content and searching capabilities. *Charlest. Advis.* 11 (1), 5–18.

Sayers, E.W., Barrett, T., Benson, D.A., Bolton, E., Bryant, S.H., Canese, K., et al., 2012. Database resources of the National Center for Biotechnology Information. *Nucleic Acids Res.* 40 (D1), D13–D25. http://dx.doi.org/10.1093/nar/gkr1184.

Schenck, R.J., Zapiecki, K.R., 2014. Back to the Future: CAS and the Shape of Chemical Information To Come. In L.R., McEwan & R.E., Buntrock (Eds.), *The Future of the History of Chemical Information* (pp. 149–58). Washington, DC: American Chemical Society.

Shultz, M., 2007. Comparing test searches in PubMed and Google Scholar. *J. Med. Libr. Assoc.* 95 (4), 442–5.

Solla, L., White, M. (Eds.), 2013. *Chemical Information for Chemists: A Primer*. Cambridge, UK: Royal Society of Chemistry.

Science information literacy and the role of academic librarians

> Have you ever noticed, when you teach, that the moment you start sharing a personal story with your students, they instantly snap to attention? You understand the value of stories. But some teachers don't insert many stories into their lessons, because they're worried that they don't have gripping stories to tell, or that they aren't good story tellers. So maybe it's worth identifying which kinds of stories are effective in making ideas stick. The answer is this: virtually any kind.
>
> *Chip Heath & Dan Heath* (Made to Stick: Why Some Ideas Survive and Others Die, *2010*)

6.1 Is there a future for information literacy instruction?

With the number of reference questions decreasing, most journals being available on-line, and database interfaces becoming more user-friendly, what is left for subject/liaison librarians to do? As information, research, and education are becoming increasingly digital, academic libraries are forced to redefine their role in supporting education and research. Teaching information literacy is a major responsibility for subject/liaison librarians, but advances in information retrieval systems, such as improved natural language searching and Semantic Web, could change their role in this area. This new environment is particularly challenging for librarians who were trained to provide traditional library services.

A Delphi study based on a survey of 13 information literacy experts looked at the possible changes in information literacy and the role of librarians in it in the next ten years (Saunders, 2009). The study developed three possible scenarios for the future of library instruction services and tried to answer the following questions:

> How prevalent will information literacy programs be within the higher education curriculum? Will academic librarians and library organizations play a significant role in the instruction and assessment of information literacy skills? If so, in what area(s) will they concentrate? Lastly, will their role be diminished as teaching faculty take on more of the responsibility for integrating this instruction into their own curricula?

The first scenario adhered to the "status quo," in which the situation remains as it is now. According to the second scenario, teaching faculty will take over instruction and assessment of information literacy, a development that will leave librarians marginalized. The third scenario envisioned a close collaboration between faculty and librarians in sharing information-literacy responsibilities. Most respondents to this survey showed optimism about the future of information literacy in academia and believed that librarians will continue to have a role to play through collaboration with faculty. The possibility that librarians could be replaced under certain circumstances has not been excluded, though, mainly because the improved and more intuitive information retrieval systems could make learning many information literacy skills unnecessary.

Departments have unique cultures, but there is also a specific culture in every university. An article discussed the importance and the difficulty of creating "a pervasively collaborative culture required by information literacy programs" and recognized that organizational culture plays a role in "campus readiness for information literacy" (Bennett, 2007). Another article suggested that librarians should avoid sticking to "a library-centric program and set an information literacy path that is relevant and valuable to course instructors and is aligned with the educational goals and mission of their institutions" (Brasley, 2008). It also described a possible framework for collaborations between librarians and teaching faculty that could lead to successful information literacy programs. Establishing partnerships between librarians and faculty, embedding librarians in academic units (Olivares, 2010), and providing online instruction (McMillen and Fabbi, 2010; York and Vance, 2009) will allow librarians to continue to play an important role in information literacy.

As suggested by Travis, in order to achieve integration of information literacy into the university curriculum, librarians and faculty need to investigate theories of change. He examined the change agency theory as a tool and a process for integrating information literacy into the general education curriculum (Travis, 2008).

The major obstacles for librarians to overcome in preserving their dominant role on the information literacy front would be faculty attitudes, lack of subject expertise, lack of technical skills, and a constantly changing dynamic environment that requires reskilling and lifelong learning (Brewerton, 2012). Librarians need to prove that their contribution to education is valuable. The number of instruction sessions is not a realistic measure of student learning, and librarians are still struggling to find a better way for assessing the impact of their efforts. By demonstrating improvement in student learning as a direct result from their instruction, librarians would be better able to justify their instructional programs.

6.2   The many faces of information literacy

The lack of consensus on how to define information literacy is at the root of the problems confronting librarians. Discipline and organizational cultures play a role in how information literacy is understood and valued (Saunders, 2009). In the future, teaching how to search and access information, which is the currently predominant model for library instruction, will become less important. If information literacy is to survive as a concept, it would need to include areas that until now either have not been supported by librarians or are the result of recent developments in technology, research, and scientific publishing. Moving away from teaching information retrieval skills in favor of providing training on managing scientific information and research data will be a great opportunity for librarians to continue to be important players in the field of information literacy. Incorporating data literacy, bibliographic management, scientific writing, and ethics of scientific communication under the umbrella of information literacy will allow librarians to find new important roles in supporting education and research in their organizations. The next sections of this chapter show how bibliographic management programs were successfully integrated in information literacy programs. The different formats (face-to-face sessions and online tutorials) and tools (LibGuides and SurveyMonkey) were used to make the teaching of information literacy and assessment of student learning more efficient.

6.3 Managing citations

Bibliographic management tools have been widely used by researchers and students to import, store, organize, and manage references that they can later use when writing research papers, theses, dissertations, journal articles, and other publications. As shown in the next sections, incorporating them into information literacy classes was very beneficial to students.

6.3.1   What bibliographic management programs allow us to do

- Easy storage of references found online
- Discovery of new articles and resources
- Automated article recommendations
- Sharing of references with peers
- Finding out who's reading what you're reading
- Storing and searching of PDFs
- Capturing references
- Inserting citations from an individual's library into a paper
- Viewing from anywhere
- Viewing saved references, along with the PDFs, on web, desktop, and mobile applications
- Taking notes and annotating articles in your library
- Automatic extraction of metadata from PDF papers
- Back-up and synchronization across multiple computers and with an online account
- PDF viewer with sticky notes, text highlighting, and full-screen reading
- Full-text search across papers
- Smart filtering, tagging, and automatic PDF file renaming
- Shared groups to collaboratively tag and annotate research papers
- Public groups to share reading lists
- Social networking features
- Usage-based readership statistics about papers, authors, and publications
- Smartphone apps
- Inserting citations from your library in a document you are writing

6.3.2 *Most popular bibliographic management programs*

The number of reference management tools is growing, and in order for users to decide which tool is best for them, they need to take into consideration many factors. Many websites and articles provide technical specifications and comparisons for these programs that help users choose the best tool for their specific needs (Duarte-Garcia, 2007; Fenner et al., 2014; McKinney, 2013; Zhang, 2012). Sometimes, one tool can be used for one purpose and another one for a different purpose. Quite often, choosing a bibliographic management tool is often a matter of personal preference.

Some of the most popular bibliographic management programs are listed below:

- CiteULike (www.citeulike.org)
- Colwiz (www.colwiz.com)
- EndNote Online (www.myendnoteweb.com)
- EndNote Desktop (www.endnote.com)
- Mendeley (www.mendeley.com)
- Papers (www.papersapp.com)
- ReadCube (www.readcube.com)
- RefWorks (www.refworks.com)
- Zotero (www.zotero.org)

EndNote, Mendeley, and Zotero are the most widely used bibliographic management programs in academic institutions, and they are discussed in more detail below.

6.3.2.1 *EndNote (Thomson Reuters)*

www.endnote.com

EndNote Desktop is a commercial reference management software package used to manage bibliographies and references when writing articles, books, theses, and other documents.

www.myendnoteweb.com

EndNote Online (previously, EndNote Web) is a free web version of EndNote (Duarte-Garcia, 2007; McKinney, 2013; Thomson Reuters, 2014; Zhang, 2012). Users can synchronize their EndNote Online account with the desktop version.

6.3.2.2 *Mendeley (Elsevier, Inc.)*

www.mendeley.com

Mendeley is a desktop and web program for managing and sharing research papers, discovering research data, and collaborating online (Habib, 2014; Haustein, 2014; Zhang, 2012). It combines Mendeley Desktop, a PDF and reference management application (available for Windows, Mac, and Linux), with Mendeley Web, an online network for researchers. Mendeley requires the user to store all data on its servers. Upon registration, Mendeley provides the user with 1000 MB of free space, which is upgradeable at a cost.

6.3.2.3   Zotero

www.zotero.org

Zotero is free, open-source software for managing bibliographic data and related research materials (e.g. PDFs). Its features include web browser integration (with Firefox), online syncing, and citing while writing. Zotero Standalone allows Zotero to be run as an independent program outside Firefox. You can add everything to Zotero— PDFs, images, videos, and snapshots of web pages.

6.3.3   Choosing a bibliographic management program

While it is important to know each tool's strengths and weaknesses, this is not the only consideration that should influence your decision. There are many websites and articles comparing the functionalities of these tools (Fenner et al., 2014; Wikipedia, 2014; Zhang, 2012), but many of these comparisons look like the technical specifications for software or hardware that you can see when looking for a digital camera or another electronic product.

For those who use Web of Science or PubMed most of the time, EndNote is probably the best tool. It is made by Thomson Reuters, which also publishes Web of Science and is optimized to work with it. EndNote provides the greatest number of citation styles (more than 5000), but this could be an advantage over the other bibliographic management tools only if you need access to many different citation styles. EndNote or Mendeley will be better options than Zotero when using other browsers than Firefox, because Zotero works best as a Firefox extension. Those who want to use EndNote Desktop have to purchase the software and install it on their computers.

Mendeley is designed to be an academic networking tool as well as a platform-independent citation management tool that syncs your data across all your computers. It would be the best choice if sharing with a network of people and finding out what citations other people are compiling in their libraries is important to you. Students tend to prefer programs that look like the social media sites they are using, and the creators of bibliographic management tools are listening, adapting to the needs and preferences of these new users. Mendeley's more "modern" interface emulates the experience they have with other social networking sites.

The more innovative feature of Mendeley, which distinguishes it from other bibliographic management programs, is that it aggregates and displays all users' citations so that users can search or browse across the entire set of citations to find resources related to their research and then add them to their own citation library for further customization. Although Mendeley maintains users' privacy, this feature may cause concerns among researchers involved in competitive areas of science who do not like their information-gathering habits to be monitored by a company.

Zotero and Mendeley allow the capture of a screenshot of a web page as well as other data about it. Syncing citations to an online Zotero account is easier and works more smoothly than syncing EndNote Desktop with EndNote Online. Zotero's capture function works with more resources (databases, catalogs, and websites) than the import function of Mendeley or the capture feature of EndNote. Users can also import citations from sites such as Amazon and Flickr. Zotero makes it easy to create tags and write notes assigned to citations.

Very often, people choose a particular bibliographic management tool because they find it easy to use. As discussed in an article, there are also generational preferences with regard to which programs people are choosing (Hull et al., 2008).

6.4   Designing information literacy instruction

The introduction of LibGuides by Springshare (www.springshare.com) in the last few years has allowed libraries to promote their information resources in a new way using multimedia and social networking tools. LibGuides are much more flexible to use than web pages controlled by rigid rules and other external (institutional) factors. For the last several years, I have been using LibGuides in teaching scientific information and bibliographic management tools (Baykoucheva, 2011). For every course in which I taught such classes, I prepared a page in a LibGuide (http://lib.guides.umd.edu/chemistryresources), in which individual course pages were listed under a tab called "Course materials." Such pages were used by students to access all resources taught in class. The assignment for the class (SurveyMonkey was used for this), a detailed handout providing essential details about search strategies, and other class-related information were also posted on the LibGuide page.

Integrating bibliographic management programs in science literacy classes allows students to learn how to do two important things: (1) perform literature and chemical property searches efficiently and (2) use a bibliographic management program to store references and cite them while writing. Classes were held in undergraduate and graduate courses in chemistry, biochemistry, molecular biology, and nutrition, as well as in a Professional Writing Program and in several honors programs.

Figure 6.1 shows a LibGuide page (http://lib.guides.umd.edu/chem272) created for a large chemistry course (CHEM272) for nonmajors with 454 students, divided into 22 sections. Teaching assistants (TAs) were trained to teach the instruction sessions for each section and grade the online assignments that students had to complete. In one hour, in addition to learning how to search literature databases, students also acquired basic knowledge of how to import references from databases into EndNote and insert citations (Cite While You Write) from their EndNote libraries into documents they were writing. Besides learning how to find and filter literature, students explored several chemistry property databases, drew molecular structures, searched for chemical compounds that corresponded to these structures, and found chemical properties and reactions for these compounds.

EndNote was used in these classes only as an example of a bibliographic management program. Once they have learned how such a program works, students can decide whether to use it or choose another tool. Zotero, Mendeley, and other new tools are free and easy to use.

6.5   How do we know we are helping students learn?

Each student had to complete an online assignment in SurveyMonkey. The assignment was graded and the grade was part of the overall grade of the student for the course. All students submitted their assignments and were very successful in answering the questions. It was very important that attendance and assignment were mandatory, which made the information

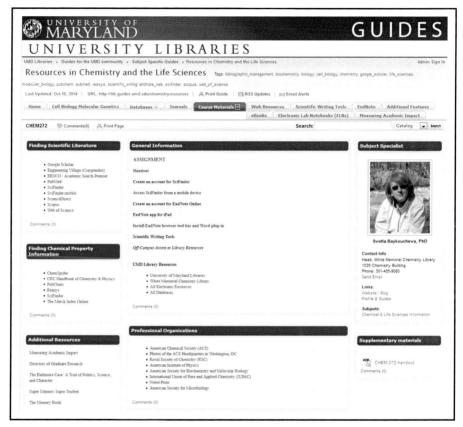

Figure 6.1 A page in a LibGuide used for teaching information literacy in an undergraduate chemistry course. The 456 students enrolled in the course were divided into 22 sections. The grades were part of the overall grade of the students for the course.

literacy class part of the whole course. As many students have acknowledged, the assignment enforced what they had learned in class and enabled them to practice with the resources and the bibliographic management program on their own. The fact that the assignment was graded and that the grade was part of their overall grade for the course was very important, as this motivated students to do it and do it as well as they could. This has substantially increased the significance of the library instruction in the eyes of both students and faculty. Now, all instructors with whom I have collaborated in incorporating information literacy classes in their courses always ask me to include an assignment and to cover EndNote.

One of the questions in the assignment required students to rank five of the resources taught in class (#1 being the most useful one). These resources were EndNote, PubMed, Reaxys, SciFinder, and Web of Science. As Figure 6.2 shows, 142 students (31%) have ranked EndNote as the most useful resource; Reaxys was ranked as the most useful resource by 163 students (36%). These results show that students realized that a bibliographic management program is very useful to them. After the successful experiment with this 200-level course, we are looking into rolling out a similar information-literacy class in a freshman chemistry course with 800 students.

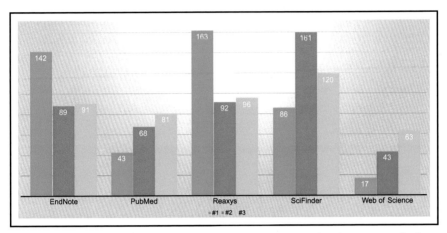

Figure 6.2 Ranking of resources by students in an undergraduate chemistry course. Students were asked to rank the three resources they had found most useful from the following ones that were taught in class: EndNote Web, PubMed, Reaxys, SciFinder, and Web of Science. EndNote was ranked as the most useful resource by 142 students (31%). The resource that was found most useful by the majority of students was Reaxys, with 163 students (36%) giving it this ranking. There were 456 students in the course, divided into 22 sections.

Students in a graduate chemistry course had to answer a similar question in their assignment. As Figure 6.3 shows, there were some differences in the preferences of students. While EndNote had similar ranking (33% of students selected it as the most useful resource), the scientific database that graduate students ranked as the most useful one was SciFinder, with 38% of students giving it this ranking.

The students in the undergraduate course chose Reaxys as the most useful resource. It is difficult to explain these differences, but it would be interesting to explore the reasons for them. It could be that the students have made these choices based on their immediate needs (e.g. other course assignments and what material their course was covering). It was interesting, though, to see how students perceived what they were taught and see them try to figure out which resources would benefit them most. It was not a surprise that students in a chemistry course ranked the chemistry databases SciFinder and Reaxys as the most important ones. What is really interesting is that they gave a bibliographic management program such a high ranking. This proves that bibliographic management should be an important component of information literacy and that students need to be exposed to these tools as early as possible.

6.5.1 What usage statistics tell us

The statistics provided by the LibGuide mentioned above show that the peaks in its usage coincided with the classes and assignments. As shown in Figure 6.4, the LibGuide was accessed 19,118 times in the year 2014. The two peaks of use (in April and September) coincided with classes that were taught in a large chemistry course (CHEM272) with 456 students, who had to access all resources through the

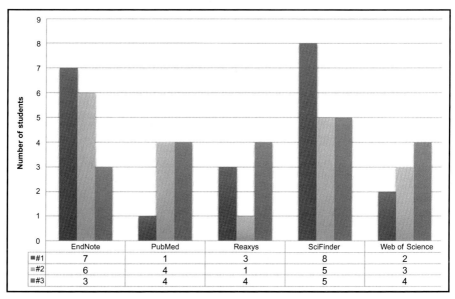

Figure 6.3 Ranking of resources by students in a graduate chemistry course (October 2014). EndNote was ranked as the #1 (most useful) resource by 33% of students and was second only to SciFinder in this respect.

Figure 6.4 Usage data for a LibGuide (http://lib.guides.umd.edu/chemistryresources) used by the author to teach scientific information and bibliographic management in chemistry and other courses at the University of Maryland College Park. As shown in the graph, the LibGuide was accessed 5546 times during the month of September 2014.

LibGuide. Particularly remarkable are the results for the month of September, when the LibGuide was used 5674 times. In the course of 12 months (from June 2013 to May 2014), the highest usage of the same LibGuide happened in October 2013 and April 2014 (Figure 6.5), which also coincided with classes and assignments. When the LibGuide stats were compared with the usage stats for SciFinder (Figure 6.6),

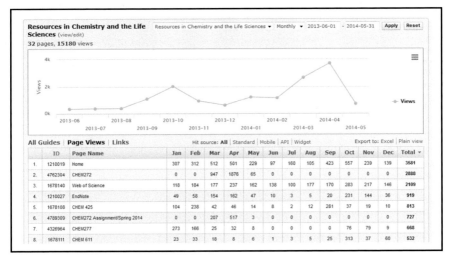

Figure 6.5 Usage statistics for a LibGuide used in teaching scientific information and bibliographic management in courses at the University of Maryland College Park during the period from June 1, 2013, to May 31, 2014.

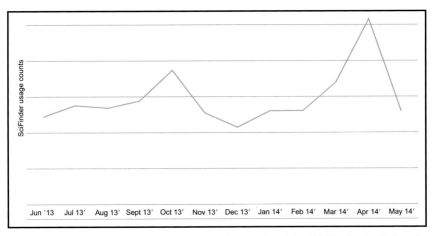

Figure 6.6 The peaks in SciFinder usage in October 2013 and April 2014 for the University of Maryland College Park coincided with the highest peaks in the access to a LibGuide used for teaching.

it was found that the times when the LibGuide was used most coincided with the highest usage of SciFinder, one of the databases students had to use in class and for their assignments.

Having statistics of this kind to present can benefit librarians who are trying to find more accurate metrics than the number of classes taught, to prove their value. Results like these could help them demonstrate that their teaching is having an impact on the use of library resources and, consequently, on the information literacy of students.

As Figure 6.4 shows, the LibGuide also provided information about which particular areas were most often used by students. In this particular case, the highest usage was registered for the course page of the large chemistry course (CHEM272), from which students accessed all databases, the assignment, and the handout, along with other materials posted for the course.

6.6 Assessing student learning

For many years, I have been using SurveyMonkey for online assignments in the courses I have taught. Having all assignments in SurveyMonkey has allowed me to keep hundreds of assignments from which I was able to reuse questions and go back and use the results for reports, papers, and adjustment of teaching strategy. I was also able to see the learning outcomes of the whole group of students, as well as the individual results. The assignments were graded, and the grades were included in the overall grade of the student for the whole course. The number of classes and their duration differed from course to course—from 50 minutes to three hours duration and from one class in a course to three classes in consecutive weeks. The assignments included from 10 questions (for undergraduate courses) to 20 questions (for graduate courses). Figures 6.7–6.10 show some examples of the questions and how the students, as a group, responded to them.

6.7 Instruction formats

There is a discussion among librarians about what formats would best suit students, instructors, and researchers in supporting information literacy. The main question that is asked is whether face-to face (F2F), entirely online instruction, or a blended format (both F2F and online) would be the most productive and efficient. As the results from surveys in my classes have consistently shown, the format students most preferred was the F2F format, as they found it easier to follow instructions in class and liked to be able to ask questions. Attending these classes was mandatory, and having an assignment that was graded motivated students to learn as much as possible in class, so that they could later do the assignment.

In the assignment, students had to answer a question about the preferred format of information literacy classes. More than 65% preferred F2F instruction; around 30%

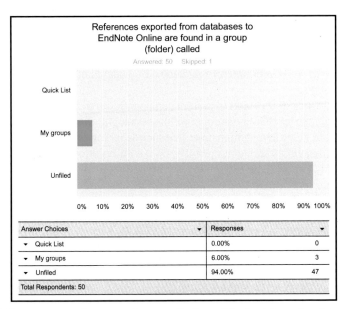

References exported from databases to EndNote Online are found in a group (folder) called

Answered: 50   Skipped: 1

Answer Choices	Responses	
▾ Quick List	0.00%	0
▾ My groups	6.00%	3
▾ Unfiled	94.00%	47
Total Respondents: 50		

Figure 6.7 Screen capture from an assignment (in SurveyMonkey) for a chemistry course. Students had to say where references exported directly from databases are found in EndNote. The correct answer is "Unfiled."

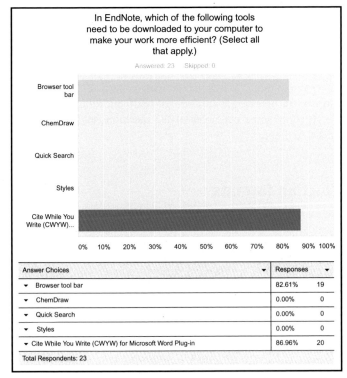

In EndNote, which of the following tools need to be downloaded to your computer to make your work more efficient? (Select all that apply.)

Answered: 23   Skipped: 0

Answer Choices	Responses	
▾ Browser tool bar	82.61%	19
▾ ChemDraw	0.00%	0
▾ Quick Search	0.00%	0
▾ Styles	0.00%	0
▾ Cite While You Write (CWYW) for Microsoft Word Plug-in	86.96%	20
Total Respondents: 23		

Figure 6.8 Screen capture from an assignment (in SurveyMonkey) that students in a chemistry course had to submit. There are two correct answers: "Browser toolbar" and "Word Plug-in."

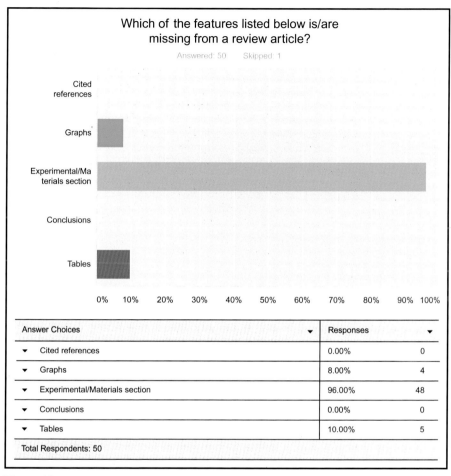

Figure 6.9 Assignment question about the difference between a research paper and a review paper. The correct answer is "Experimental/Materials section."

preferred blended instruction, and only a small percentage wanted completely online instruction. As shown in Figure 6.11, students in a chemistry course (CHEM277) for chemistry majors ranked the in-class presentation as most useful, followed by the online handout prepared for the class.

Responses from students in another course about the preferred format for information literacy instruction showed that face-to-face instruction was the preferred format (67%), followed by "face-to-face and online tutorial" (42%) (Figure 6.12). Only 4% of students preferred the online only format. The question was included in an assignment for a course (UNIV100, Integrated Life Sciences) in the University of Maryland College Park.

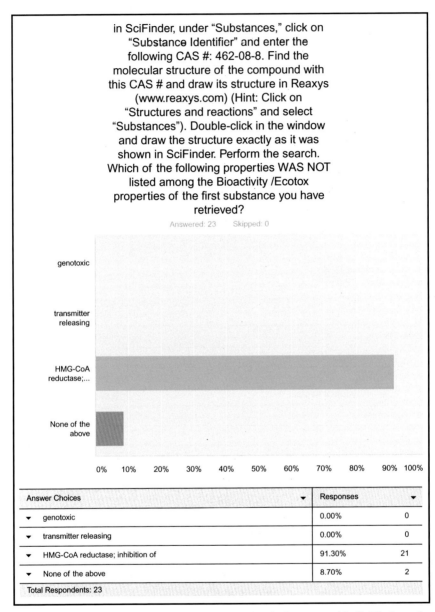

Figure 6.10 Students were asked to (1) search for a chemical compound in SciFinder, using its CAS number; (2) find the molecular structure of that compound and draw its structure in another database, Reaxys; (3) look at the bioactivity/ecotoxicity properties of the first retrieved compound; and (4) choose which of the properties listed under the question was not associated with this compound. The correct answer is "HMG-CoA reductase, inhibition of."

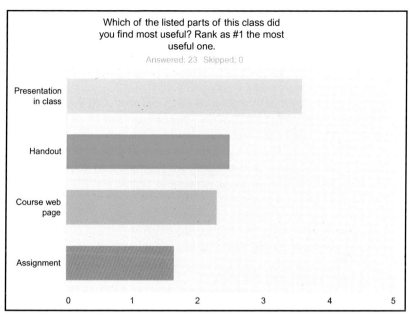

Figure 6.11 Responses from students in a 200-level chemistry course for chemistry majors to a question about which component of the information literacy instruction they found most useful. The question was included in an assignment that students had to complete in February 2014.

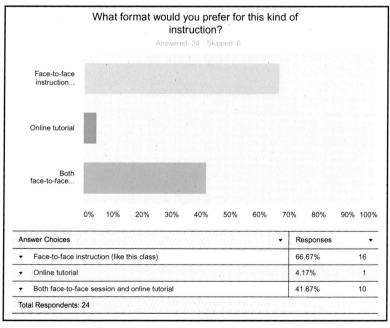

Answer Choices	Responses	
Face-to-face instruction (like this class)	66.67%	16
Online tutorial	4.17%	1
Both face-to-face session and online tutorial	41.67%	10
Total Respondents: 24		

Figure 6.12 Responses from students about preferred format for information literacy instruction. The question was included in an assignment for a course (UNIV100, Integrated Life Sciences) in the University of Maryland College Park. Face-to-face instruction was the most preferred format (67%), followed by "face-to-face and online tutorial" (42%). Only 4% of students preferred online (only) format.

6.8   Other elements of information literacy

As the importance of research data is increasing, eScience has emerged as a new opportunity for academic libraries to support research and education in their institutions.
Helping researchers manage their data and integrating data literacy in library instruction will be the main areas of engagement for librarians in the future (Qin and
D'Ignazio, 2010). The advancement of data literacy will depend on identifying and
developing core competencies and standards that can serve as a framework for its
inclusion in libraries' information literacy programs (Prado and Marzal, 2013; Vogt,
2013). The roles of librarians in supporting eScience, in general, and data literacy, in
particular, is covered in more detail in Chapter 8.

Academic librarians now discuss in their information literacy classes other areas
such as scientific communication, Open Access, new forms of publishing, ethics of
scientific publishing, scientific impact, social networks, and altmetrics. These topics
are covered in other chapters of this book.

All assignment examples are from chemistry courses taught at the University of
Maryland College Park. Screen captures from assignments are reproduced with permission from SurveyMonkey.

6.9   Sample questions for assignments in science courses

1. Adding an asterisk (*) at the end of a word would allow you to retrieve:
   - More articles
   - Fewer articles
   - Specific articles
2. After performing a search in a database, which of the following actions will reduce the
   number of results? (Select all that apply.)
   - Selecting a range of publication year(s)
   - Adding additional keywords
   - Choosing a particular document type (e.g. review)
   - All of the above
3. Which of the following actions will allow you to expand your search?
   - Putting an asterisk (*) after a word
   - Using specific terms
   - Using broader terms
   - Using a larger number of keywords
   - All of the above
   - None of the above
4. Which of the features listed below is/are missing from a review article?
   - Cited references
   - Graphs
   - Experimental/Materials section
   - Conclusions
   - Tables

5. In PubMed, (1) click on the "Advanced" option; (2) type ENZYME INHIBITORS in the first box. (3) Select "MESH Major Topic" from the pull-down menu on the left of this box; (4) click on "Show index list" on the right of the box; (5) double-click on "Enzyme inhibitors." (6) In the second box, type CHOLESTEROL; (7) from the pull-down menu on the left of this box, select "MESH Term." (8) Click on "Show index list" on the right of the box; (9) double-click on "Cholesterol"; (10) perform the search and limit it by time period (Hint: select "Custom") to the time period from 1/1/2009 to 12/31/2010. Select "Review" under "Article types." How many documents were retrieved for this time period?
   - 111
   - 246
   - 18
   - 308
   - None of the above

6. In Web of Science, enter "aspirin and cancer" (no quotation marks) as a topic. In the box under "Refine Results," type "treatment" and click on the magnifier on the right of the box; on the next screen, check the box next to "Oncology" under "Web of Science categories," and click on "Refine." Narrow down the results by publication year, from 2009 to 2012 (Hint: make the years chronological by changing "Record count" to "Alphabetical"). How many documents were retrieved for this time period? Enter the answer in the box.
   - 204
   - 15
   - 43
   - 120
   - None of the above

7. After performing a search in SciFinder, which of the following actions is important to perform to narrow down the number of results (select all that apply)?
   - Refine the list by publication year
   - Remove duplicates
   - Add additional keywords
   - Limit to a particular document type (e.g. review)
   - All of the above

8. Perform a search for properties of aspirin in SciFinder (Hint: Go to "Substance identifier"; then on the next screen, click on the aspirin CAS number.) What is the melting point (experimental) for aspirin (in degrees Celsius)?
   - 135
   - 138
   - 145
   - 164
   - None of the above

9. In SciFinder, perform a search on a topic (use "Explore References") and type "enzyme inhibitors AND cholesterol" (without the quotation marks) in the search box. From the options you were presented with, which one would be best to choose?
   - References were found containing "enzyme inhibitors and cholesterol" as entered.
   - References were found containing both of the concepts "enzyme inhibitors" and "cholesterol."
   - References were found containing either the concept "enzyme inhibitors" or the concept "cholesterol."
   - References were found containing the concept "enzyme inhibitors."
   - References were found containing the concept "cholesterol."

10. In SciFinder, under "Substances," click on "Substance Identifier" and enter the following CAS #: 462-08-8. Find the molecular structure of the compound with this CAS # and draw its structure in Reaxys (www.reaxys.com) (Hint: Click on "Substances, Names, Formulas" and select "Substances"). Double-click in the query editor window on the left of the screen and draw the structure exactly as it was shown in SciFinder. (If you have a problem opening the drawing window (it is Java-based), click on "Structure editor" at the bottom of this window and select "Dotmatics Elemental.") After drawing the structure, click on "Transfer Query" and click on "Search." Which of the following properties *was not* listed among the bioactivity properties of the first substance you have retrieved?
    - Genotoxic
    - Transmitter-releasing
    - HMG-CoA reductase, inhibition of
    - None of the above

11. In Reaxys, click on "Substances, Names & Formulas," select "Substances," and type the following CAS # in the appropriate box: 504-24-5. This CAS # corresponds to which of the following chemical compounds?
    - Cholesterol
    - Toluene
    - Arachidonic acid
    - 4-Aminopyridine
    - None of the above

12. Use the Advanced Search option in ChemSpider; check the box next to "Select by Properties." Select "Molecular Formula" and enter the following empirical formula: $C33H35FN2O5$. To which of the following substances does this formula correspond to?
    - Lovastatin
    - Lisinopril
    - Crestor
    - Atorvastatin

13. In PubChem, find property information for 3-aminopyridine. Which of the following numbers corresponds to the molecular weight (in g/mol) of this compound as listed under "Computed Properties" in the "Chemical and Physical Properties" section?
    - 96.12908
    - 94.12356
    - 94.11454
    - 94.23901

14. Perform a search in Scopus using the following string of search terms: "reverse transcriptase and HIV and hepatitis B" (no quotation marks). Limit the results to the year 2010. Which source title (journal) has published the highest number of articles on this topic in this particular year?
    - The Journal of Immunology
    - AIDS
    - Antiviral Therapy
    - None of the above

15. Perform a search in Scopus using "hplc fatty acids" as search terms (without the quotation marks). How many review articles were published on this topic in the year 2009?
    - 83
    - 121
    - 51
    - 14
    - 8
    - None of the above

16. Which of the resources shown in class were most useful to you? The most useful would be "1."
    - EndNote
    - PubMed
    - Reaxys
    - SciFinder
    - Scopus
    - Web of Science

17. Select from the list below the resources about which you have learned for the first time in this class.
    - PubChem
    - PubMed
    - SciFinder
    - Reaxys
    - Scopus
    - Web of Science

18. References imported directly from databases to EndNote Web will be placed in
    - Quick List
    - Unfiled
    - Trash
    - None of the above

19. Perform a search in Scopus using the following keywords: "plant physiology and drought" (no quotation marks). Refine your results by adding "fungal" as an additional topic. Limit your results to those published in the year 2010. How many articles were published in the "Journal of Plant Physiology" for that year?
    - 203
    - 16
    - 8
    - 37
    - None of the above

20. Perform a search in Agricola on the EBSCO platform. Type "plant physiology" in the first box and "drought" in the second box (no quotation marks). Refine your search to SCHOLARLY ARTICLES and to the time period 2002–2006. How many articles were published during this time period?
    - 402
    - 389
    - 206
    - 14
    - None of the above

21. If you could choose the format of this kind of instruction, which of the following options would you prefer?
- Face-to-face instruction with online tutorials (like this class)
- Online tutorial only

22. Did you find this class and the assignment useful and why? Was the assignment difficult to do and how long did it take you to do it?

References

Baykoucheva, S., 2011. Using Campus Guides for leveraging Web 2.0 technologies and promoting the chemistry and life sciences information resources. In: Abstracts of Papers, 241st ACS National Meeting & Exposition, Anaheim, CA, United States, March 27–31, 2011, CINF-50.

Bennett, S., 2007. Campus cultures fostering information literacy. *portal Libr. Acad.* 7 (2), 147–67.

Brasley, S.S., 2008. Effective librarian and discipline faculty collaboration models for integrating information literacy into the fabric of an academic institution. *New Dir. Teach. Learn.* 2008 (114), 71–88.

Brewerton, A., 2012. Re-skilling for research: investigating the needs of researchers and how library staff can best support them. *New Rev. Acad. Librarian.* 18 (1), 96–110. http://dx.doi.org/10.1080/13614533.2012.665718.

Duarte-García, E., 2007. Personal managers of bibliographic reference data bases: characteristics and comparative analysis. *Prof. Inf.* 16 (6), 647–56. http://dx.doi.org/10.3145/epi.2007.nov.12.

Fenner, M., Scheliga, K., Bartling, S., 2014. Reference management. *Opening Science.* Retrieved from http://book.openingscience.org/tools/reference_management.html.

Habib, M., 2014. Mendeley Readership Statistics available in Scopus. Retrieved from http://blog.scopus.com/posts/mendeley-readership-statistics-available-in-scopus.

Haustein, S., 2014. Tweets and Mendeley readers: two different types of article level metrics. Retrieved July 2, 2014, from http://www.slideshare.net/StefanieHaustein/haustein-ape2014-30482551.

Hull, D., Pettifer, S.R., Kell, D.B., 2008. Defrosting the Digital Library: Bibliographic Tools for the Next Generation Web. *PLoS Comput Biol* 4 (10), e1000204. http://dx.doi.org/10.1371/journal.pcbi.1000204.

McKinney, A., 2013. EndNote Web: web-based bibliographic management. *JERML* 10 (4), 185–92. http://dx.doi.org/10.1080/15424065.2013.847693.

McMillen, P., Fabbi, J., 2010. How to be an E³ librarian. *Public Serv. Q.* 6 (2–3), 174–86.

Olivares, O., 2010. The sufficiently embedded librarian: defining and establishing productive librarian-faculty partnerships in academic libraries. *Public Serv. Q.* 6 (2–3), 140–9.

Prado, J.C., Marzal, M.A., 2013. Incorporating data literacy into information literacy programs: core competencies and contents. *Libri* 63 (2), 123–34.

Qin, J., D'Ignazio, J., 2010. The central role of metadata in a science data literacy course. *J. Libr. Metadata* 10 (2–3), 188–204.

Saunders, L., 2009. The future of information literacy in academic libraries: a Delphi study. *portal Libr. Acad.* 9 (1), 99–114.

Thomson Reuters, 2014. EndNote. Retrieved from http://www.endnoteweb.com.

Travis, T.A., 2008. Librarians as agents of change: working with curriculum committees using change agency theory. *New Dir. Teach. Learn.* 2008 (114), 17–33.

Vogt, L., 2013. eScience and the need for data standards in the life sciences: in pursuit of objectivity rather than truth. *Syst. Biodivers.* 11 (3), 257–270.

Wikipedia, 2014. Comparison of reference management software. Retrieved from https:// en.wikipedia.org/wiki/Comparison_of_reference_management_software.

York, A.C., Vance, J.M., 2009. Taking library instruction into the online classroom: best practices for embedded librarians. *J. Libr. Adm.* 49 (1/2), 197–209. http://dx.doi. org/10.1080/01930820802312995.

Zhang, Y., 2012. Comparison of select reference management tools. *Med. Ref. Serv. Q.* 31 (1), 45–60. http://dx.doi.org/10.1080/02763869.2012.641841.

Information literacy and social media: interview with Chérifa Boukacem-Zeghmouri

7

Chérifa Boukacem-Zeghmouri

Chérifa Boukacem-Zeghmouri is a lecturer in information and communication science at Claude Bernard Lyon 1 University, a research university. She is a research member of the Equipe de recherche de Lyon en sciences de l'Information et de la Communication—ELICO Research Laboratory. She is cohead of "Urfist de Lyon," a regional service dedicated to training researchers in finding and using scientific information. Her PhD in information science (defended in 2004) was dedicated to an economic analysis of the transition of the academic library services to an electronic environment. From 2006 to 2010, she chaired a national project on the use of electronic resources in the French academic context, applying a socioeconomic approach. She led the first national study on the return on investment (RoI) of electronic resources in the French academic network, funded by Lyon 1 University and Elsevier, and the first Algerian study on the use of electronic resources, funded by Springer. Since 2008, she has been involved in organizing national and international scientific events and has published many articles in French (*Bulletin des Bibliothèques de France, Documentaliste—Sciences de l'Information, Enjeux de l'Information et de la Communication*) and international (*Learned Publishing, Canadian Journal of Information and Library Science, Serials, ILDS, Scientometrics*) journals.

Svetla Baykoucheva: You are involved in providing information support for researchers and graduate students. Could you describe the organization you are part of and what you do? Which are the disciplines and geographic areas that you are providing support for?

Chérifa Boukacem-Zeghmouri: Urfist (Unité Régionale de Formation à l'Information Scientifique et Technique (ULRIST), 2014) are services devoted to scientific information for the academic community. Seven Urfists exist in France, each one overseeing a different geographical region. These services were created by the Ministry of Higher Education and Research 30 years ago, with the mission to help academics master digital scientific information. As public service institutions, the Urfists are devoted to academic communities and provide free support and training. The information monitoring tools that are developed within the Urfist network are also free to use for academics.

Urfist of Lyon (URFIST de Lyon, 2014) was the first one to be created, in 1980, at a time when the computerization of society was in full swing. The Lyon Urfist oversees the regions of Auvergne, Burgundy, and Rhône-Alpes. As soon as it was created, the Lyon Urfist started helping academics appraising new tools and new thematic, which were all highly relevant in the digital scientific world. Thirty years later, the Urfist's mission hasn't fundamentally changed, but focuses more on researchers (from universities, Centre National de la Recherche Scientifique (CNRS), Institut National de la Recherche Agronomique (INRA), Institut National de la Santé et de la Recherche Médicale (INSERM), etc.) and graduate students, the latter seen as young researchers in training. Such a focus was decided by the ministry in the hope of supporting research communities as they face the transitions in scientific communication and publishing.

The Lyon Urfist is administratively linked to the Claude Bernard Lyon 1 University, which acts as a host institution. The service operates on a specific fund allocated by the ministry. With this fund, the Urfist covers its costs (missions, equipment, the payment of outside trainers, etc.). Salaries are not included in this package. The Lyon Urfist team is composed of a lecturer in information science, who is affiliated to Lyon 1 and to a research laboratory (ELICO) and who is in charge of sciences, technology, and medicine (STM)-related fields. The team's librarian is in charge of the humanities and social sciences (HSS) fields. Finally, the team also includes an administrator. A wide range of STM and HSS disciplines are covered by the Urfist team, and outside experts are called in for specific missions. Very diverse topics can therefore be studied within each field, making the Urfist's list of contacts a valuable resource.

SB: What kind of training do you provide?

CBZ: The training programs offered by the Lyon Urfist are designed to adapt to the needs of researchers. In our semester catalog of training programs, we offer short programs (3–4 h) and longer sessions (1 or 2 days). Thanks to these programs, researchers can manipulate tools (such as databases—Inspec, Web of Science (WoS)) or management tools for bibliographic references (such as Mendeley and Zotero) and expand their critical understanding of current issues (publishing in Open Access journals, for example). They can also develop their skills for scientific writing. The catalog of services is based on the needs voiced by researchers. Training sessions are led by the Urfist team or by outside experts. Sessions take place in small groups and consist of presentations, discussions, questions, and practical exercises.

Apart from the training catalog, we also organize information and current awareness seminars that focus on a current issue (e.g. predatory publishing, article publishing charges, megajournals). These seminars take place during lunch hours or at the end of the workday. They are useful as they focus on a particular issue; they also trigger discussions between researchers who come from different institutions and various academic fields.

We organize some seminars and study days on a more periodical basis. These events are centered on themes that unite key participants in research, libraries, and scientific publishing, and they are very popular among our followers.

We also meet researchers for individual training sessions. These services are "tailor-made" and can range from consulting (what are the best sources or tools to carry out a specific mission?), to customized training (ResearchGate, ResearcherID), to simple consultations (around questions such as copyright). Researchers often call us in urgent situations, when they are applying for grants or positions. They appreciate our quick response time, our flexibility, and the precision of our answers.

We help research laboratories handle services and tools on the long term, as well as guide them to better understand the underlying challenges of certain issues. Our goal is to help them be more aware of the current competitive academic context in which they work. We first carry out a review before creating the research team's training program. This review helps to define the needs and levels of training of all team members. Requests to train research teams are sometimes issued following an evaluation made by the National Commission for the Assessment of Research.

SB: You have done surveys of graduate students and established researchers about their use of science resources. How would you assess the science information literacy of the groups you have studied?

CBZ: The most striking and all-encompassing aspect, common to all fields, consists in the "googlization" of searches on specialized databases. What I mean by this is that researchers transfer their Google habits to specialized databases like WoS. Practices therefore converge toward a simple search with a single keyword. Navigation then takes over when the researcher sorts out and selects search results.

However, we do observe differences between graduate students and established researchers regarding the selection and interpretation of what they find. Established researchers have reference points, critical thinking, and analytical standards that allow them to judge the value of an article, the quality of its source, or the keyword's relevance. These researchers can also create more complex combined searches as they go, thus filtering information and refining search results.

Graduate students don't have that type of background to situate themselves in the environment. This is why they tend to rely on their networks, developing communication skills that serve to mitigate their "googlization" habits. The rise of academic social networks greatly contributes to the amplification of this phenomenon.

SB: What are the most dominant trends in finding, managing, and using information by these groups? What are the things that surprised you in what you have found?

CBZ: The first trend consists in researchers increasingly depending on Google and Google Scholar to search for information and scientific literature. This is explained by Google's ease of use, a quality highly appreciated by researchers. Google is the starting point for any search and it presents the user with the keywords and many different facets (images, videos, news, etc.).

The second trend revolves around the fact that searching, finding, and managing information is now embedded in the whole research cycle. These tasks blend in with the actual scientific activity. Researchers can simultaneously conduct experiments, obtain initial results, and search for information in a scientific database. These tasks no longer happen in a sequential time frame but are rather integrated and subordinated to the other tasks linked to scientific activity.

SB: What differences did you find among researchers in the way they gather and use information, depending on the discipline and their status?

CBZ: When conducting field studies, I realized that each individual's research practices and information management tactics depended on his/her belonging to a structured and/or conservative field of study. These two aspects seem to be central defining elements.

The researcher's status also comes into play, as long as we also take age into account. In France, this is the age at which we attain different types of status differs, depending on the field of study. I would argue that, given the current context, the generational effect is as important as the status effect. It is also interesting to perform studies that take into account the differences in subject areas.

SB: How do established researchers and graduate students use social networks? How do they differ with respect to the discipline or their status (for example, established researchers or graduate students)?

CBZ: The study I conducted recently (Boukacem-Zeghmouri, 2014) allowed me to highlight the fact that young researchers were more familiar with social networks (public and academic) than the previous generation. In certain cases, supervisors came to discover the existence and workings of social networks thanks to their doctoral students. To a certain extent, the latter took on the role of initiator.

Doctoral students use social networks not only to build knowledge and expertise in their research areas but also to learn more about their research communities and scientific production. Social networks are used to help them select the most "reliable" articles for reading, citing, and finding out who the expert authors in their field are, so that they could follow the most interesting research teams for their careers. Academic social networks are for them a kind of expertise network. They help young researchers better navigate and situate themselves in a world where benchmarks aren't always clearly indicated. However, the study showed that doctoral students were most often consumers and observers instead of being contributors. They prefer not to expose themselves for fear of making blunders and attracting negative attention from their peers.

Between the fields of chemistry, biology, and computer sciences, the chemistry students revealed themselves to be the most conservative and cautious in their use of social networks. Computer science students were, conversely, the most innovative users.

SB: How do graduate students communicate with their supervisors and their peers?

CBZ: In the STM fields, doctoral students and their supervisors are very close. It's common to see the doctoral student's office right next to the supervisor's. Supervisors and students communicate in formal situations (during thesis planning meetings and research team meetings) and informal moments (lunch breaks, train rides, etc.), as they do lab work or travel to conventions together. Doctoral students communicate cautiously with their peers. They remain vigilant because of the competition and strain linked to searching for job positions, participating in research contracts and establishing a reputation. As is the custom in France, the first contact with peers is often done through the supervisor. But initial contacts also take place through social networks, in which students add each other to their lists of contacts.

SB: *Are there other interesting findings from your research that you would like to mention?*

CBZ: My most recent work revealed the significance of risk and reputation for STM researchers, who now operate in a fast-changing environment where previously held values are being pushed aside. Analyzing the information searching and publishing practices of researchers—while also taking into account notions of risk and reputation—allow us to understand the structural evolutions that surround new models of scholarly communication. These studies contribute to understanding current trends and anticipating those to come.

References

Boukacem-Zeghmouri, C., 2014. When information literacy meets a scholarly borderless world. Paper presented at the Facing the Future: Librarians and Information Literacy in a Changing Landscape, pp. 8-14, IFLA Satellite Meeting, Limerick, Ireland, August 14–15.

Unités Régionale de Formation à l'Information Scientifique et Technique (URFIST), 2014. Retrieved September 9, 2014, (URFIST) from http://urfistinfo.hypotheses.org/a-propos.

URFIST de Lyon, 2014. Retrieved March 23, 2015, from http://urfist.univ-lyon1.fr.

Coping with "Big Data": eScience

> The Petabyte Age is different because more is different. Kilobytes
> were stored on floppy disks. Megabytes were stored on hard disks.
> Terabytes were stored in disk arrays. Petabytes are stored in the
> cloud. As we moved along that progression, we went from the folder
> analogy to the file cabinet analogy to the library analogy to—well, at
> petabytes we ran out of organizational analogies.
>
> *(Andersen, 2008)*

8.1 Introduction

The growing interest in making research publicly available resulted in the development of web infrastructure and services that allow the production, management, and storage of large volumes of data. The rise of "Big Data," a term used to denote the exponential growth of large volumes of data, was called by Jim Gray from Microsoft "the fourth paradigm" in scientific research, the first three being experimental, theoretical, and computational sciences (Tolle et al., 2011). Initially a characteristic of some areas of biology, particle physics, astronomy, and environmental sciences, this revolution has spread to the humanities, social sciences, and other disciplines (Bartling and Friesike, 2014).

The rapid accumulation of research data is creating an urgent need for action before organizations and companies get swamped by unmanageable volumes of data. How are all these data going to be managed, preserved, and stored? Who will do it? How are we going to find them? What are researchers going to do with the data?

A government report, "U.S. Open Data Action Plan," provided directions for "managing information as a national asset and opening up its data, where possible, as a public good to advance government efficiency, improve accountability, and fuel private sector innovation, scientific discovery, and economic growth" (US Government, 2014).

8.2 Types of research data

Data are everything produced in the course of research. According to the National Science Foundation (NSF), "What constitutes such data will be determined by the community of interest through the process of peer review and program management. This may include, but is not limited to: data, publications, samples, physical collections, software and models" (US National Science Foundation (NSF), 2010). The UK government defined data as "Qualitative or quantitative statements or numbers that are assumed to be factual, and not the product of analysis or interpretation" (UK Government, 2014).

Data are generated through experimental, statistical, or other processes and are often stored locally. Some of the most common types of data include laboratory experimental data, computer simulation, textual analysis, and geospatial data. In the social sciences, data include numeric data, audio, video, and other digital content generated in the course of social studies or obtained from administrative records. For scholars in the humanities, data consist of textual and semantic elements. Data can be structured or unstructured; simple or complex; and uniform or heterogeneous in terms of syntax, structure, and semantics. They are organized in datasets that are curated, archived, stored, and made accessible.

Graphic displays can communicate information much more clearly than pure numerical data. Some of the visual representations of data may include a dynamic component such as time and computer animation and simulation. Visualization of research data is particularly important in some disciplines, such as astrophysics, biology, chemistry, geophysics, mathematics, meteorology, and medicine.

8.3   Managing data

Data management is an important part of the research process. Researchers have to ensure that their data are accurate, complete, and authentic. Some of the tools that are used to manage research data depend on the nature of the discipline and the type of experiment, while others are of a more general type that can be used in many research activities, such as data storage and analysis (Liu et al., 2014; Ray, 2014).

The management of data consists of four layers: curation, preservation, archiving, and storage.

8.3.1   Data curation

A uniform understanding of the meaning or semantics of the data and their correct interpretation requires defining a number of characteristics or attributes. The process of data authentication, archiving, management, preservation, retrieval, and representation is known as data curation. In the course of this process, metadata (which are data about data) are created to help data discovery and retrieval. Metadata summarize data content, structure, context, relationships, and relevance. They can be viewed as a specific kind of information structured to describe, explain, locate, or make it possible to retrieve, use, or manage data. There are three types of metadata:

- Descriptive metadata are about the content of data (e.g. title, author, and keywords).
- Administrative metadata stipulate preservation, ownership, and technical information about formats.
- Structural metadata are about the design and specification of data structures.

Metadata include information about data provenance, a term used to describe the history of data from the moment they were created, through all their modifications in content and location. Data provenance is particularly important to scholarship today, because the digital environment makes it so easy to copy and paste data (Frey and Bird, 2013; Liu et al., 2014).

The adoption of the digital object identifier (DOI) was an important step in improving the process of dataset identification, verification, and discovery. A DOI (discussed in more detail in Chapter 15 of this book) is a unique persistent identifier that is attached to a dataset, an article, or other creative work. It is similar to an International Standard Book Number (ISBN), which allows a book to be tagged, discovered, and retrieved.

8.3.2   Data preservation, archiving, and storage

Data preservation and storage are some of the biggest challenges that data present. Surveys of researchers from different institutions have shown that research data are often found in spreadsheets that do not allow the performance of complex analyses, sharing, or reuse. They are saved on computers or external hard drives—sometimes without any backup. More and more data, though, are now stored in the "cloud" or managed with locally installed software tools (Frey and Bird, 2013). Hardware architecture for storage is very important, but software also plays a significant role. Data can be stored as flat files, indexed files, relational files, binary files, or any other electronic format. There are some data that are not easy to preserve, which would require special software for storing them. If only some of the data are to be preserved, rules have to be designed to filter data and prevent the loss of those that are important.

8.4   Data standards

Data standards are accepted agreements on format, representation, definition, structuring, manipulation, tagging, transmission, use, and management of data. A data standard describes the required content and format in which particular types of data are to be presented and exchanged. Establishing standards is an important step toward managing data in a more consistent way, and their absence results in data corruption, loss of information, and lack of interoperability.

In the United States, the development and application of technical standards for publishing bibliographic and library applications are coordinated by the National Information Standards Organization (NISO) (www.niso.org). NISO is currently involved in the development of assessment criteria for nontraditional research outputs such as datasets, visualization, software, and other applications (National Information Standards Organization, 2015). The organization also provides webinars and training in data management, discovery tools, new forms of publishing, and alternative metrics (discussed in Chapter 14 of this book) for evaluation of research.

Building new data standards is a difficult task, because different areas of science require different standards. Genomics research, for example, generates enormous volumes of data that are not all numeric. In a review addressing the creation of data standards for the life sciences, Vogt presented arguments for establishing two different types of standard—one for scientific argumentation and another one for the production of data. The first type would "…use criteria of logical coherence and empirical grounding. They are important for the justification of an explanatory hypothesis. Data and metadata standards, on the other hand, concern the data record itself and all steps and actions taken during data production and are based on virtues of objectivity …" (Vogt, 2013).

Vogt also suggested that, "… in order to meet the requirements of eScience, the specification of the currently popular minimum information checklists should be complemented to cover four aspects: (i) content standards, which increase reproducibility and operational transparency of data production, (ii) concept standards, which increase the semantic transparency of the terms used in data records, (iii) nomenclatural standards, which provide stable and unambiguous links between the terms used and their underlying definitions or their real referents, and (iv) format standards, which increase compatibility and computer-parsability of data records."

8.5 Citing data

While there are established conventions for citing published papers, there is no uniformly accepted format for citations of digital research data. The currently emerging conventions vary by discipline, but some common elements within these conventions are becoming obvious. An article presents an overview of the citation practices of individual organizations and disciplines and identifies the following set of "first principles" for data citation that can be adapted to different disciplines, organizations, and countries to guide the development and implementation of data citation practices and protocols (CODATA-ICSTI, 2013):

1. *Status of data*: Data citations should be accorded the same importance in the scholarly record as the citation of other objects.
2. *Attribution*: Citations should facilitate giving scholarly credit and legal attribution to all parties responsible for those data.
3. *Persistence*: Citations should be as durable as the cited objects.
4. *Access*: Citations should facilitate access both to the data themselves and to such associated metadata and documentation as are necessary for both humans and machines to make informed use of the referenced data.
5. *Discovery*: Citations should support the discovery of data and their documentation.
6. *Provenance*: Citations should facilitate the establishment of provenance of data.
7. *Granularity*: Citations should support the finest-grained description necessary to identify the data.
8. *Verifiability*: Citations should contain information sufficient to identify the data unambiguously.
9. *Metadata standards*: Citations should employ widely accepted metadata standards.
10. *Flexibility*: Citation methods should be sufficiently flexible to accommodate the variant practices among communities but should not differ so much that they compromise interoperability of data across communities.

Data publication and citation are very important to the scientific community, as they make the scientific process more transparent and allow data creators to receive credit for their work. Just like citation of articles, citation of datasets demonstrates the impact of the research and will benefit authors. Developing mechanisms for peer review of data will ensure the quality of datasets and will allow analysis of data and conclusions made on the basis of these data (Kratz and Strasser, 2014).

8.6 Data sharing

Many research labs are now using cloud storage facilities such as Dropbox, OneDrive, Google Drive, and Box to share data (Mitroff, 2014). Services such as the subscription-based Basecamp (https://basecamp.com) are used by research groups and larger units for project management purposes. Scientific online collaboration platforms such as colwiz (www.colwiz.com) and Zoho (www.zoho.com) are also popular for managing research data and files. Chapter 9 of this book is entirely devoted to electronic laboratory notebooks (ELNs), which are used for recording, managing, and sharing research data and other information.

As studies of the research practices of scientists have shown, there are significant differences between the disciplines in the practices used for managing and sharing data, which often depend on the culture of the lab or the organization (Bird and Frey, 2013; Harley, 2013; Long and Schonfeld, 2013; Meadows, 2014; Ray, 2014; Tenopir et al., 2011). Researchers in some disciplines share data through a central database available to many researchers, while others do not share them until they publish the results.

An example of such specific culture of research is the area of bioinformatics, a conglomerate of interdisciplinary techniques, which transforms large volumes of diverse data into more manageable and useable information. The handling and analysis of data in this discipline are performed through annotation of macromolecular sequences and structure databases and by classifying sequences or structures into groups. Researchers can interact with these databases and manipulate them, which is of critical importance and benefits users (Scott and Parsons, 2006).

Wiley's Researcher Data Insights Survey of 90,000 recent authors of papers in different disciplines demonstrated significant differences across research fields and geographic locations in how researchers share their data (Ferguson, 2014; Wiley, 2014). The survey results showed that only 52% of researchers make their data publicly available. Of the data presented for seven countries, German scientists were those who were sharing their results most often (55%). Significant differences were found in how researchers shared their data: 67% provided their data as a supplementary material to papers submitted for publication; 37% posted them on a personal, institutional, or project web page; 26% deposited them in an institutional repository; 19% used a discipline-specific repository; and 6% submitted them to a general-purpose repository such as Dryad or figshare.

Researchers had different motivations for sharing data: 57% said that sharing was common practice in their organization; 55% wanted to make their research more visible and increase its impact; and half of the respondents considered sharing a public benefit. Other reasons included journal, institutional, or funders' requirements; transparency and reuse; discoverability and accessibility; and preservation.

According to an article that examined the differences between disciplines in sharing data, the life sciences researchers were the most sharing (66%), while those in the social sciences and the humanities were the least likely (36%) to share their data. Another article based on results from surveys and interviews with researchers on current data-sharing practices discussed the different concerns that researchers have about sharing data (Tenopir et al., 2011). According to this article, the most common concerns of researchers were associated with legal issues, data misuse, and incompatible data types. Most data are not copyrightable in the United States, and copyrights usually do not apply, but the situation is different in other countries. It is important that authors understand their rights before publishing their data and adhere to ethical rules when reusing them.

8.7 eScience/eResearch

The term eScience (also called eResearch) has different meanings, but they all are associated with large datasets and tools that allow producing and managing large volumes of digital scientific data. The National Science Board of the NSF has published a document, "Digital Research Sharing and Management," addressing the complexities of modern science in this way: "Science and engineering increasingly depend on abundant and diverse digital data. The reliance of researchers on digital data marks a transition in the conduct of research" (US National Science Foundation (NSF), 2011). The increasing scale, scope, and complexity of datasets are changing fundamentally the way that researchers share, store, and analyze data. Large and complex datasets pose significant challenges for the research community, because they are sometimes difficult to store, manage, and analyze with conventional approaches and tools. Such datasets may need to be subjected to special management processes and even to be split and stored on several servers. Another major problem in managing large volumes of data is that the existing applications often use proprietary data types, which prevents the parsing of data types from other third-party applications.

The long-term preservation of digital data is of critical importance for their future use. The US National Science Board outlined the following strategies for cooperation in data preservation and storage (US National Science Board, 2011):

> Strategic partnerships between key stakeholder communities should be developed to collectively support the development of effective data repositories and stewardship policies. Funding agencies, university-based research libraries, disciplinary societies, publishers, and research consortia should distribute responsibilities that address the establishment and maintenance of digital repositories.

8.8 Data repositories and organizations involved in data preservation

Some of the available data repositories are funded by governments or by professional organizations, while others are commercial enterprises. The discipline-specific repositories are specifically designed to accommodate metadata for the particular subject area. They are smaller and are usually funded by grants. In the United States, the federal funding for repositories is increasing, and Europe is also making significant investments in building such facilities. Some of the data repositories and organizations involved in data preservation and sharing are presented below:

The **Board on Research Data and Information** at the National Academies (The National Academies, 2014) has a mission to improve the policy and use of digital data and information in science and society.

The **Coalition for Networked Information (CNI)** (www.cni.org) publishes papers focused on eScience and data management. Also, on the CNI website (http://www.cni.org/event-calendar/), there is a list of upcoming conferences and workshops that include some on eScience/data management.

The **Committee on Data for Science and Technology (CODATA)** (www.codata.org), an interdisciplinary committee under the umbrella of the International Council for Science (ICSU), is dedicated to improving the quality, management, and availability of data in science and technology.

DataCite (datacite.org) is a data registry created and supported by several libraries from different countries. DataCite assigns DOIs—an effort that is coordinated by the California Digital Library (CDL), Purdue University Libraries, and the Office of Science and Technology (OSTI) of the Department of Energy (DOE). The creation of DataCite and the assignment of DOIs were a big operation, which turned out to be even more difficult than the creation of ISBNs.

Databib (http://www.re3data.org/) is a searchable catalog, registry, directory, and bibliography of research data repositories. It allows identifying and locating online repositories of research data.

Digital Curation Centre (DCC) (www.dcc.ac.uk) is a leading international center for data management located in the United Kingdom.

The **Data Observation Network for Earth (DataONE)** (www.dataone.org) is an aggregator supported by the NSF under the DataNet program. DataONE provides archiving facilities for ecological and environmental data produced by scientists worldwide.

The **Data Net Project (US National Science Foundation (NSF))** is an aggregator of resources for interdisciplinary research.

The **Dataverse Network** (thedata.org) provides open-source software that can be downloaded, installed, and customized by an institution or organization to host their own Dataverse repository.

The **Digital Public Library of America (DPLA)** (dp.la) is a free aggregated repository combining resources from many organizations, including the Library of Congress, HathiTrust, and the Internet Archive. It provides access to large digitized collections of books, images, historic records, and audiovisual materials.

Distributed Data Curation Center (D2C2) (d2c2.lib.purdue.edu) is a research center at Purdue University.

The **Dryad Digital Repository** (datadryad.org) is a nonprofit free repository for data underlying publications from the international scientific and medical literature. It is a curated resource that accepts many data types and makes them discoverable.

DuraSpace (www.duraspace.org) is an independent nonprofit organization founded in 2009 through a collaboration of the Fedora Commons organization and the DSpace Foundation, which are two of the largest providers of open-source repository software. Many institutions worldwide that use DSpace or Fedora open-source repository software belong to the DuraSpace community.

figshare (figshare.com) is a commercial online digital repository where researchers can preserve and share their research outputs, including figures, datasets, images, and videos. It is free to upload content and free to access, in adherence to the principle of open data. It allows users to upload any file format that can be visualized in a browser (Fenner, 2012).

The **Harvard Dataverse Network** (thedata.harvard.edu/dvn) is open to all researchers worldwide to publish research data across all disciplines. It is a repository for long-term preservation of research data that provides permanent identifiers for datasets.

The **National Information Standards Organization (NISO)** (www.niso.org) is a nonprofit organization that supports the discovery, retrieval, management, and preservation of published content. It connects libraries, publishers, and information systems vendors that develop technical standards and provide education about technological advances in information exchange.

*Open*DOAR (www.opendoar.org) is a directory of open-access academic repositories. It is one of the SHERPA services including RoMEO and JULIET that are discussed in Chapter 2 of this book.

The **Registry of Research Data Repositories** (www.re3data.org) offers researchers, funding organizations, libraries, and publishers a directory of existing international repositories for research data. It was created initially by the Berlin School of Library and Information Science at the Humboldt-Universität zu Berlin, the Library and Information Services (LIS) Department of the GFZ German Research Centre for Geosciences, and the KIT Library at the Karlsruhe Institute of Technology (KIT). All records from Databib are now integrated in it, and by the end of 2015, re3data.org will become an imprint of DataCite and be included in its suite of services.

VIVO (www.vivoweb.org) is an interdisciplinary network that enables collaboration among researchers across all disciplines and allows users to search for information on people, departments, courses, grants, and publications. Initiated by Cornell University (vivo.cornell.edu), Vivo is an open-source semantic web application, which is installed locally. Participating institutions have control over content and over the search and browse capabilities.

Zenodo (zenodo.org), a repository hosted at CERN (near Geneva, Switzerland), was created through the OpenAIREplus project of the European Commission to enable researchers and institutions in the European Union (EU) to share multidisciplinary research results such as data and publications that are not part of the other institutional or disciplinary repositories.

8.9   Data management plans

Research institutions and government organizations have become concerned about how researchers manage, preserve, and share their data. NSF and other government funding organizations have introduced new policies aimed at preserving the integrity of data and allowing their sharing, analysis, and discovery (Peters and Dryden, 2011;

US National Science Foundation, 2014). Grant proposals submitted to the NSF now must include a data management plan, which describes how the proposal will conform to NSF policy on the dissemination and sharing of research results (US National Science Foundation, 2013). This plan may include the following information (US National Science Foundation, 2014):

- Types of data, samples, physical collections, software, curriculum materials, and other materials to be produced in the course of the project.
- Standards to be used for data and metadata format and content (where existing standards are absent or deemed inadequate, this should be documented along with any proposed solutions or remedies).
- Policies for accessing and sharing including provisions for appropriate protection of privacy, confidentiality, security, intellectual property, or other rights or requirements.
- Policies and provisions for reuse, redistribution, and the production of derivatives.
- Plans for archiving data, samples, and other research products and for preservation of access to them.

The requirements do not stipulate how data will be managed internally by investigators, but rather focus on how they will be shared and disseminated externally and preserved in the long term. Each data management plan addresses the following aspects of research data: metadata (descriptions of created materials such as types of data, samples, physical collections, and software), standards (for data and metadata formats and content), access policies, and archiving.

8.10   eScience and academic libraries

It is rare to see a job announcement for a science librarian today that does not have "data management" or "eScience" mentioned in the title. Academic librarians are exploring new roles in supporting researchers by helping them in the acquisition, storage, retrieval, analysis, and preservation of data (Bailey, 2011; Bennett and Nicholson, 2007; Carlson et al., 2011; Corrall et al., 2013; Cox and Corrall, 2013; Heidorn, 2011; Prado and Marzal, 2013; Wang, 2013).

In 2011, the Association of Research Libraries (ARL) (Association of Research Libraries, 2014) offered a 6-month-long program called eScience Institute, designed to help research libraries develop a strategic agenda for eScience support, with a particular focus on the sciences. The institute included a series of modules for teams to complete at their institutions and organized a final workshop for participants in the program (Antell et al., 2014).

8.10.1   Data curation

Academic libraries will have an important role to play in ensuring the archiving and preservation of data (Ball, 2013; Cox and Pinfield, 2014; Garritano and Carlson, 2009). Datasets will be treated as important collections. Several years ago, I interviewed James Mullins, Dean of Purdue University Libraries, who was one of the pioneers in introducing eScience in academic libraries (Baykoucheva,

2011). This is how Dr. Mullins described the possible roles that librarians could play in eScience:

> Working in the area of data management draws upon the principles of library and archival sciences. Our ability to see structure to overlay on a mass of disparate "parts," as well as the ability to identify taxonomies to create a defined language for accessing and retrieving data is what is needed from us ...

> Once a librarian has the experience of talking with researchers about their research and the challenges they have with managing data, it becomes clear that the most important factor is not our subject expertise (although some subject understanding is needed) but rather the librarian's knowledge of metadata and taxonomies. In the old days we would have said that this is "cataloging and classification," but today, to convey that we have morphed into a new role, it is best to use the more technical terminologies since it may help identify our "new" role as a cutting edge initiative and not be encumbered with past misperceptions.

8.10.2   Data preservation and storage

Should libraries be responsible for storing research data? And if they are not, who should be doing it? Many universities now maintain institutional repositories, which are online archives for collecting, preserving, and disseminating the intellectual output of a research institution. These repositories provide a preservation and archival infrastructure that allow researchers to share and get recognition for their data. The university repositories have started accepting datasets, but libraries will have to adapt the existing standards for bibliographic metadata to such new research outputs as datasets. How do you catalog entities that have no title or publisher, sometimes even no obvious author? The new generation of institutional repository systems maintained by academic libraries will need to be able to handle a variety of metadata formats, allow immediate analysis, and provide streaming capabilities. More sophisticated archival and preservation features will be essential in making the repositories more attractive to researchers for depositing their works.

8.10.3   Data literacy

Being able to read text is not a problem, but how about reading and understanding data? To be able to do this, you need to have certain skills in data literacy, which is a different type of literacy. There is a need for creating data literacy standards and core data literacy competencies, and some efforts in this direction have already been made (Prado and Marzal, 2013).

Research institutions and individual academic departments are now looking at introducing a mandatory requirement that graduate students attend training in data management. Until now, graduate students obtained some data management information as part of research ethics training. Academic libraries could find a niche opportunity in offering such training to graduate students through workshops or by integrating it in the library instruction classes they teach in science courses.

8.10.4   Are the academic libraries promising more than they can deliver?

The initial view of how libraries could support eScience was that they would be helping researchers draft the data management plans required by funding agencies (Ball, 2013). It soon became clear that researchers do not need librarians' help, as these plans tend to be standard for any given discipline, and there are templates that can be used to prepare them. As for helping researchers create metadata for their datasets, this role might not be so easy to play, mainly because of subject expertise. A possible involvement for libraries could be to educate students and researchers in how to create metadata, but it is still not clear how important this role could be, as most datasets have very few of these.

Steven Bell considers the engagement of libraries with "Big Data" a "double-edged sword." He sees "… a place for academic librarians to help their institutions succeed in the effort to capture and capitalize on Big Data. That is well and good, but let's remember to bring to it the same sensibility that has made us wise evaluators and collectors of information in the past" (Bell, 2013). In spite of all the enthusiasm about eScience, some authors are not certain what role librarians will play in it. An article that looked at the skills and requirements for eScience-related library positions concluded that at present, eScience librarianship is not a well-defined field and that "the role of librarians in e-science is nebulous" (Alvaro et al., 2011).

In his interview included in Chapter 10 of this book, Gary Wiggins discussed the role of academic libraries in supporting eScience, cautioning that this might be an effort for which libraries will need more time to prepare. The biggest problems that academic librarians will face will be their lack of subject expertise and specific skills and gaining the trust of researchers as experts in this field. As suggested in an article, "librarians need to act as a voice for balance and reason within their organizations. While Big Data analytics has many interesting possibilities that should be explored, there is no substitute for more traditional methods of research" (Hoy, 2014).

8.11   Conclusion

Data constitute an integral part of the research life cycle, and where they are not present to support published results questions may be raised about the trustworthiness of those results. Many scientific journals now require raw data to be included in papers submitted for publication. Research data, alone, have no meaning, unless they are properly described and interpreted. This is how Joshua Schimel summarized the transition from raw data to interpretation (Schimel, 2011):

> The role of scientists is to collect data and transform them into understanding. Their role as authors is to present that understanding. However, going from data to understanding is a multi-step process. The raw data that come from an instrument need to be converted to information, which is then transformed into knowledge, which in turn is synthesized and used to produce understanding.

The issue with the huge volumes of data is not only having enough space for it but also describing or indexing of datasets or creating metadata, which take much longer.

In the future, efforts in the area of data management will be focused on preserving the integrity of data and their quality. Data classification, visualization, and contextualization will be equally important.

With research becoming more interdisciplinary and global, the need to design technologies and platforms that will allow researchers to collaborate more efficiently will become even more important. Further implementation of a semantic-based eScience infrastructure, Science 2.0 tools, and new web technologies will make information sharing and collaboration much easier (Bird and Frey, 2013). As the mandate by the NSF and other funding agencies forces a change in the sharing practices, researchers will see more benefits of making their results more open.

To create a culture of data citation and linking on a larger scale, significant changes are needed in the publishing infrastructure that will make data citation, linking, and reuse an integral part of the publication models. Proper citing and attribution of data will make it more difficult for someone to "steal" research data. The adoption of stricter rules for data attribution is likely to convince more researchers to share their data with their peers and even to make them openly available.

Academic libraries are transitioning from providing platforms for information to providing data-related services (Arms et al., 2009; Gradmann, 2014; Peters and Dryden, 2011; Tenopir et al., 2014). Engaging in these new areas will provide academic libraries with an opportunity to reimage themselves and become more closely aligned with the research and educational missions of their institutions.

References

Alvaro, E., Brooks, H., Ham, M., Poegel, S., Rosencrans, S., 2011. e-Science librarianship: field undefined. *Issues in Science & Technology Librarianship* 66.

Andersen, C., 2008. The End of Theory: The Data Deluge Makes the Scientific Method Obsolete. *Wired Magazine*, June 27. Retrieved December 11, 2014, from http://archive.wired.com/science/discoveries/magazine/16-07/pb_theory.

Antell, K., Foote, J.B., Turner, J., Shults, B., 2014. Dealing with data: science librarians' participation in data management at association of research libraries institutions. *Coll. Res. Libr.* 75 (4), 557–74. http://dx.doi.org/10.5860/crl.75.4.557.

Arms, W.Y., Calimlim, M., Walle, L., 2009. EScience in practice: lessons from the Cornell Web Lab. *D-Lib Magazine* 15 (5–6).

Association of Research Libraries, 2014. E-Research. Retrieved December 9, 2014, from http://www.arl.org/focus-areas/e-research#.VIcj13uansk.

Bailey, C.W. Jr., 2011. E-science and Academic Libraries Bibliography, http://www.digital-scholarship.org/ealb/ealb.htm.

Ball, J., 2013. Research data management for libraries: getting started. *Insights* 26 (3), 256–60. http://dx.doi.org/10.1629/2048-7754.70.

Bartling, S., Friesike, S., 2014. *Opening Science: The Evolving Guide on How the Internet Is Changing Research, Collaboration and Scholarly Publishing.* Springer International Publishing: Heidelberg, New York, NY.

Baykoucheva, S., 2011. What do libraries have to do with e-Science? An interview with James L. Mullins, Dean of Purdue University Libraries. *Chem. Inf. Bull.* 63 (1), 45–9. http://acsinf.org/content/what-do-libraries-have-do-e-science.

Bell, S., 2013. Promise and problems of Big Data: from the Bell Tower. *Libr. J.* (March 13). Accessed on May 16, 2015. http://lj.libraryjournal.com/2013/03/opinion/steven-bell/promise-and-problems-of-big-data-from-the-bell-tower/#_.

Bennett, T.B., Nicholson, S.Q., 2007. Connecting users to numeric and spatial resources. *Soc. Sci. Comput. Rev.* 25 (3), 302–318.

Bird, C.L., Frey, J.G., 2013. Chemical information matters: an e-Research perspective on information and data sharing in the chemical sciences. *Chem. Soc. Rev.* 42 (16), 6754–6776. http://dx.doi.org/10.1039/c3cs60050e.

Carlson, J., Fosmire, M., Miller, C.C., Nelson, M.S., 2011. Determining data information literacy needs: a study of students and research faculty. *portal: Libr. Acad.* 11 (2), 629–57. http://dx.doi.org/10.1353/pla.2011.0022.

CODATA-ICSTI, 2013. Out of cite, out of mind: the current state of practice, policy, and technology for the citation of data. *Data Sci. J.* 12, CIDCR1–CIDCR75.

Corrall, S., Kennan, M.A., Afzal, W., 2013. Bibliometrics and research data management services: emerging trends in library support for research. *Libr. Trends* 61 (3), 636–74.

Cox, A.M., Corrall, S., 2013. Evolving academic library specialties. *J. Am. Soc. Inf. Sci. Technol.* 64 (8), 1526–42. http://dx.doi.org/10.1002/asi.22847.

Cox, A.M., Pinfield, S., 2014. Research data management and libraries: current activities and future priorities. *JOLIS* 46 (4), 299–316.

Fenner, M., 2012. Figshare: Interview with Mark Hahnel. *Gobbledygook*. Retrieved 5 July 2014, from http://blogs.plos.org/mfenner/2012/02/16/figshare-interview-with-mark-hahnel.

Ferguson, L., 2014. How and why researchers share data (and why they don't). Retrieved from http://exchanges.wiley.com/blog/2014/11/03/how-and-why-researchers-share-data-and-why-they-dont/.

Frey, J.G., Bird, C.L., 2013. Cheminformatics and the semantic web: adding value with linked data and enhanced provenance. *Wiley Interdiscip. Rev. Comput. Mol. Sci.* 3 (5), 465–81. http://dx.doi.org/10.1002/wcms.1127.

Garritano, J.R., Carlson, J.R., 2009. A subject librarian's guide to collaborating on e-science projects. *Issues in Science & Technology Librarianship* (Spring). http://dx.doi.org/10.5062/F42B8VZ3.

Gradmann, S., 2014. From containers to content to context: the changing role of libraries in eScience and eScholarship. *J. Doc.* 70 (2), 241–60.

Harley, D., 2013. Scholarly communication: cultural contexts, evolving models. *Science* 342 (6154), 80–82. http://dx.doi.org/10.1126/science.1243622.

Heidorn, P.B., 2011. The emerging role of libraries in data curation and E-science. *J. Libr. Adm.* 51 (7–8), 662–72. http://dx.doi.org/10.1080/01930826.2011.601269.

Hoy, M.B., 2014. Big Data: an introduction for librarians. *Med. Ref. Serv. Q.* 33 (3), 320–26. http://dx.doi.org/10.1080/02763869.2014.925709.

Kratz, J., Strasser, C., 2014. Data publication consensus and controversies. *F1000Research* 3, 94. http://dx.doi.org/10.12688/f1000research.3979.2.

Liu, Q., Bai, Q., Giugni, S., Williamson, D., Taylor, J.E., 2014. *Data Provenance and Data Management in eScience*. Springer: Heidelberg, New York.

Long, M.P., Schonfeld, R.C., 2013. Supporting the Changing Research Practices of Chemists. Research Support Services: Chemistry Project. Retrieved from http://www.sr.ithaka.org/research-publications/supporting-changing-research-practices-chemists.

Meadows, A., 2014. To share or not to share? That is the (research data) question…. Retrieved December 11, 2014, from http://scholarlykitchen.sspnet.org/2014/11/11/to-share-or-not-to-share-that-is-the-research-data-question/.

Mitroff, S., 2014. OneDrive, Dropbox, Google Drive, and Box: Which cloud storage service is right for you? Retrieved December 10, 2014, from http://www.cnet.com/how-to/onedrive-dropbox-google-drive-and-box-which-cloud-storage-service-is-right-for-you.

National Science Board, 2011. Digital research data sharing and management. Retrieved December 11, 2014, from http://www.nsf.gov/nsb/publications/2011/nsb01211.pdf.

National Science Foundation, 2013. Grant Proposal Guide, Chapter II—Proposal Preparation Instructions. Retrieved December 6, 2014, from http://www.nsf.gov/pubs/policydocs/pappguide/nsf13001/gpg_2.jsp#dmp.

National Science Foundation, 2014. NSF Data Management Plan Requirements. Retrieved December 11, 2014, from http://www.nsf.gov/eng/general/dmp.jsp.

National Information Standards Organization, 2015. NISO Alternative Assessment Metrics (Altmetrics) Initiative. Retrieved January 8, 2015, from http://www.niso.org/topics/tl/altmetrics_initiative.

National Science Foundation (NSF), 2010. Data Management & Sharing Frequently Asked Questions (FAQs). Retrieved July 30, 2014, from http://www.nsf.gov/bfa/dias/policy/dmp-faqs.jsp.

National Science Foundation (NSF), 2011. Digital Research Sharing and Management, Retrieved July 30, 2014, from http://www.nsf.gov/nsb/publications/2011/nsb1124.pdf.

Peters, C., Dryden, A.R., 2011. Assessing the academic library's role in campus-wide research data management: a first step at the University of Houston. Sci. Technol. Libr. 30 (4), 387–403. http://dx.doi.org/10.1080/0194262X.2011.626340.

Prado, J.C., Marzal, M.A., 2013. Incorporating data literacy into information literacy programs: core competencies and contents. Libri 63 (2), 123–34.

Ray, J.M. (Ed.), 2014. Research Data Management: Practical Strategies for Information Professionals, Purdue University Press: West Lafayette, Indiana.

Schimel, J., 2011. Writing Science: How to Write Papers That Get Cited and Proposals That Get Funded. Oxford University Press: New York.

Scott, R.K., Parsons, A., 2006. Information management—biodata in life sciences. In: Ekins, S. (Ed.), Computer Applications in Pharmaceutical Research and Development. John Wiley & Sons, Inc.: Hoboken, NJ, pp. 167–186.

Tenopir, C., Palmer, C.L., Metzer, L., van der Hoeven, J., Malone, J., 2011. Sharing data: practices, barriers, and incentives. Proc. Am. Soc. Inf. Sci. Technol. 48 (1), 1–4.

Tenopir, C., Sandusky, R.J., Allard, S., Birch, B., 2014. Research data management services in academic research libraries and perceptions of librarians. Libr. Inf. Sci. Res. 36 (2), 84–90.

The National Academies, 2014. Board on Research Data and Information (BRDI). Retrieved December 29, 2014, from http://sites.nationalacademies.org/PGA/brdi/index.htm.

Tolle, K.M., Tansley, D.S.W., Hey, A.J.G., 2011. The fourth Paradigm: Data-intensive scientific discovery. Proceedings of the IEEE 99 (8), 1334–7.

UK Government, 2014. Open Data White Paper: Unleashing the potential. Retrieved December 11, 2014, from http://data.gov.uk/sites/default/files/Open_data_White_Paper.pdf.

US Government, 2014. U.S. Open Data Action Plan. Retrieved July 30, 2014, from http://www.whitehouse.gov/sites/default/files/microsites/ostp/us_open_data_action_plan.pdf.

Vogt, L., 2013. eScience and the need for data standards in the life sciences: in pursuit of objectivity rather than truth. Systematics and Biodiversity 11 (3), 257–70.

Wang, M., 2013. Supporting the research process through expanded library data services. Program 47 (3), 282–303. http://dx.doi.org/10.1108/PROG-04-2012-0010.

Wiley, 2014. Researcher Data Sharing Insights. Retrieved December 11, 2014, from https://scholarlykitchen.files.wordpress.com/2014/11/researcher-data-insights-infographic-final.pdf.

Managing research data: electronic laboratory notebooks (ELNs)

9

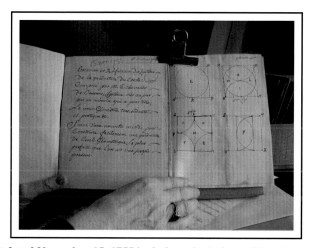

A manuscript dated November 15, 1755 includes calculations of the characteristics of the circle. (From the Archives of the Institut de France.)

9.1 Introduction

In 1985, Margot O'Toole, a new postdoc in the laboratory of Thereza Imanishi-Kari at the Massachusetts Institute of Technology (MIT), was asked to do experiments that would expand a previous work carried out in the lab and reported in a 1986 paper published in the journal *Cell*. Her inability to make the procedures described in the paper work in her experiments and the discrepancies she discovered between the laboratory notebooks and the published results made her suspect that there was something wrong with the experimental setup and the data reported in the paper.

Many scientists can easily relate to a situation like this one, as it often happens that a newcomer is asked to use protocols established in the lab and reproduce results that have already been reported in a paper. By making her suspicions public, O'Toole stirred controversy and vigorous discussion about fraud in science and triggered what became to be known as "The Baltimore Case" (Kevles, 2000). The most famous of the authors of the paper in the center of this storm, David Baltimore, was awarded the Nobel Prize in Medicine in 1975 for his discovery of retroviruses and the way they reproduce themselves. Baltimore subsequently became interested in studying the immune system. The *Cell* paper had reported results that could lead to the induction of genetic modifications of the immune system.

Managing Scientific Information and Research Data

It was concluded that the paper did not contain fraudulent data and the authors were exonerated, but the investigation took a significant toll on all who were involved in it. Baltimore was the recently appointed president of the prestigious Rockefeller University; a year later, he resigned, exhausted by fighting a long battle over these accusations of fraud. He later (1997 to 2006) served as president of the California Institute of Technology (Caltech). What became clear from the investigations and the Congress hearings, though, was that the notebooks in the lab had not been kept in a very orderly and consistent manner.

Recent surveys of researchers with regard to how they record and maintain their data have shown that data are often not backed up and are usually stored on computer hard drives. Once the paper is published, some of the data cannot be recovered or found, which makes it difficult to reproduce the results if there are doubts about the authenticity of the experimental data. Many scientific journals now require raw experimental data to be included in the papers submitted for publication and use special software to detect suspicious results and plagiarism (Enserink, 2012; Marusic, 2010).

9.2   Recording research data

Conducting experiments and documenting the results are daily routine for scientists. Accurate and traceable documentation enables other scientists to reproduce procedures and results for building on this experience and to confirm the validity of reported results. Laboratory notebooks are used to record each step in conducting an experiment and processing data. Originally, these notebooks were paper-based. Computerization of research systems and practices increased the need for electronic laboratory notebooks (ELNs) that allow using sophisticated computational features and reliable electronic documentation.

Recording the provenance of data is an important feature of ELNs that allows this information to be traced, so that it can later be queried for further analysis. The ELNs record data provenance; document the inputs, entities, systems, and processes that influence data of interest; and allow the work flow of the preparation, execution, evaluation, interpretation, and archiving of research data to be traced. They provide a historical record of the data and their origins. Thus, generated evidence increases the reliability of research results, keeps the research process transparent to individual research collaborators, and is essential for auditing and compliance analysis of research data.

9.3   Paper *vs* digital

Paper notebooks are portable and do not require power supply. It is quick and easy to write notes in them and record data obtained at the lab bench. The disadvantages are the risk of loss or damage. They are also not searchable and cannot be shared, some handwriting might not be readable, and language issues could prevent the information recorded from being interpreted correctly. One constant complaint from lab supervisors is that they find it very difficult to chase students' notebooks. After someone leaves the lab, it is sometimes impossible to reproduce the experimental procedures or obtain the same results as those already published in a paper. Anyone who has tried to follow the descriptions in a Materials and Methods section of a paper knows that it is very possible for some important details to have been omitted.

As science is getting much more data-intensive, interdisciplinary, and collaborative, the paper notebook cannot satisfy the needs of modern scientists. Surveys of researchers have shown that some of the most frequently mentioned challenges in the lab are finding data and information when they are needed; storing and organizing data; sharing data with others; using too many systems and databases; and just keeping up with the growing volume of data. The ELN is an alternative to the paper notebook. It is a system that allows the creation, storage, retrieval, and sharing of data and other laboratory information. The first generation of ELNs mirrored a paper notebook and served as its replacement. The models available on the market today are much more sophisticated, with features and capabilities that greatly increase productivity, save time, and provide protection for intellectual property. They can also be searched, shared, and accessed from anywhere.

Widely implemented in industry, ELNs have not been readily accepted in the academic environment. One of the reasons for this is that research in academia is decentralized and scientists have more freedom to choose how they organize their labs and perform their work. Another obstacle lies in the existing technology—the ELNs available today are far from being user-friendly, or cheap, and they lack certain features that would have satisfied some specific needs. The fear that a generation of models will soon make those available today obsolete and that the information stored in the current ELNs might not be transferable is also contributing to the indecisiveness of academic researchers in embracing this technology.

9.4　Finding information about ELNs

Judged by the growing number of articles, conferences, and websites devoted to ELNs, this technology is attracting the attention of researchers, lab managers, tech companies, investors, and science administrators who are trying to make research more efficient and cost-effective. Professional societies such as the American Chemical Society have been devoting entire technical sessions to ELNs at their national meetings. The papers presented at these sessions discuss many different aspects of the technology and its implementation in research labs. Several recent reviews follow the evolution of the ELN and describe the current types and models available today and the challenges involved in switching from paper to digital (Bird et al., 2013; Machina and Wild, 2013a; Pearsall, 2013; Rudolphi and Goossen, 2011; Taylor, 2011).

9.5   Benefits of using ELNs

The most important benefits of using ELNs are summarized below:

- All kinds of research data and other laboratory data can be captured, organized, and pre-served, in many different formats.
- Experiments can be cloned to reduce redundancy in record keeping and save time.
- Scientific data and observations are easier to search, find, and use.
- Records can be accessed at any time, from anywhere, especially on mobile devices.
- Errors and inefficiencies are eliminated.
- Scientific fraud is prevented and intellectual property protected (E-Signatures).
- Sharing and collaboration is easy (virtual teams).
- Access permissions can be managed by role.
- Interoperability (e.g. the direct import of data from instruments).
- Back history of records can be tracked.
- Can be customized to satisfy specific needs.

9.6   Types of ELNs

The term "ELN" means different things to different people. There are simple note-taking tools, discipline-specific models, and blog-based ELNs; proprietary or open-source; local or cloud-based (Elliott, 2010; Frey and Bird, 2011; Lagoze et al., 2012; Rubacha et al., 2011). The first generation of ELNs was created in the 1990s, and it originated from Big Pharma. The second generation was developed in the 2000s and became more popular among university researchers. Programs such as Evernote and Dropbox were web-based platforms and were all-purpose tools. Some more lab-oriented ELNs also came out at that time and allowed the sharing of data. eCaA, LabArchives, and Ruro were among them. The third generation was started in 2011–2012, when the University of Wisconsin began a pilot project, exploring the interests of researchers in this technology.

9.6.1   Basic ELNs

These models are designed for basic data capture and are the closest alternative to paper notebooks. An example of a basic note-taking tool is Evernote. It allows the organization and tagging of information in different formats such as text, web, images, and e-mail (Elliott, 2012; Walsh and Cho, 2013). An article presented the experience of a translational science laboratory at the New York University School of Medicine (NYUSoM) with a pilot project designed to determine if an ELN could effectively replace paper documentation in an academic research setting (Walsh and Cho, 2013). All routine experimental information was recorded on a daily basis over a period of six months. Students working in research laboratories were also surveyed to find out what they thought of the advantages and limitations of ELNs and paper notebooks. Most of the notebook users reported the inability to freehand into a notebook as a limitation when using ELNs. The authors found that the many advantages provided by Evernote significantly outweighed the lack of freehand capability and that using it as an ELN effectively replaced paper notebooks in an academic research setting and provided many other advantages compared to traditional paper notebooks.

Another popular note-taking tool is *Microsoft SharePoint*, a tool that allows the user to organize and share files with others. *The IPython Notebook* is a web-based interactive environment that allows text, mathematics, plots, and rich media to be combined into a single document. The notebooks are normal files that can be converted to other formats and shared with others. In *GoogleDrive*, users can create Google Docs and store more than 30 other file types that can be shared from any device.

9.6.2   Universal/multidisciplinary ELNs

Multidisciplinary or universal ELNs could find wider acceptance than the models that are specifically tailored toward particular disciplines or for specific purposes (John et al., 2011; Voegele et al., 2013).

LabArchives offer an ELN for research and a classroom edition that can be used in university courses (Figures 9.1–9.3). Some of the features of LabArchives are listed below:

- Built-in viewers for common scientific files.
- Link to specific versions of data (e.g. protocols).
- Extensible platform using widgets (e.g. for buffer calculations and searching PubMed).
- Automatic uploading of data from automated instruments.
- Storing of all versions of all data.
- Direct interface with Microsoft Office.

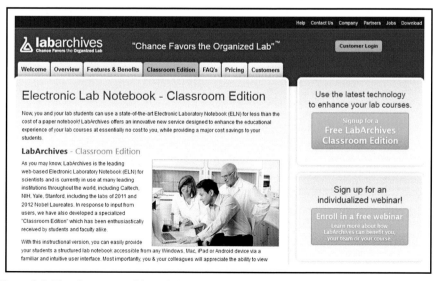

Figure 9.1 The classroom edition of the LabArchives ELN.

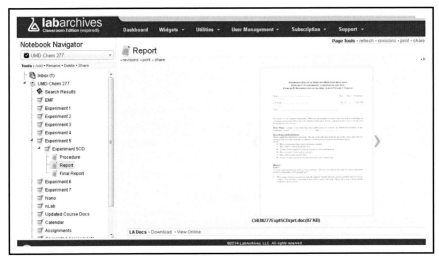

Figure 9.2 Interface of the classroom edition of the LabArchives ELN.

Figure 9.3 Screen capture of a LabArchives ELN interface showing widget for buffer preparation.

All screen captures from LabArchives are reproduced with permission from LabArchives.

An article discussed the implementation of the LabArchives ELN at Yale University and the future plans for promoting this technology, in collaboration with other academic institutions and ELN vendors (Bogdan, 2014).

9.6.3 Specific ELNs

ELNs for specific needs are designed for life sciences and chemistry laboratories in industry, academic institutions, and hospitals (Beato et al., 2011; Christ et al., 2012; Elliott, 2013; Frey and Bird, 2013; John et al., 2011; Lagoze et al., 2012; Oleksik et al., 2014). Many pharmaceutical companies have implemented ELNs or have created their own ELNs for drug discovery (Machina and Wild, 2013b). Some ELNs were built to serve the needs of chemists (Christ et al., 2012; Frey and Bird, 2013; Lagoze et al., 2012; MacNeil, 2011; Milsted et al., 2013; Rajarao and Weiss, 2011; Rudolphi and Goossen, 2011).

The ELN from *Perkin Elmer* allows the recording of in vitro and in vivo experiments. The company has put a lot of emphasis on providing features, such as a drawing tool (ChemDraw), that are important to chemists. It also allows adding custom configurations suitable for chemistry and biology.

Accelrys Notebook is a multidiscipline ELN suitable for biology and chemistry research. The cloud version provides valuable features such as drawing tools; reaction visualization and stoichiometric calculations; single and multistep reactions; and access to reactant, reagent, and solvent databases. More advanced features, such as integration with existing chemical databases and inventory for quick lookup of reactants and reagents, are available when the ELN is hosted locally. *Agilent Technologies* developed an integrated framework that includes chromatography data systems, an ELN, and a scientific data management system.

9.6.4 Open-source ELNs

The web makes it possible for scholars to expand their data beyond text and share images, video, and other kinds of materials. Open-source models are very attractive, as they have the benefit of being free. A universal open-source ELN was described in a recently published article (Voegele et al., 2013). Another example of an open-source and blog-based model is LabTrove, which was created by a group of academic researchers at the University of Southampton in the United Kingdom (Bird et al., 2013; Frey and Bird, 2013; Milsted et al., 2013). It represents a hybrid solution that includes the use of both a paper notebook and an electronic one.

9.7 Introducing ELNs in academic institutions

Some universities are trying to experiment with ELNs for research and education. In 2012, the University of Wisconsin-Madison carried out a pilot project and tested ELNs in several labs to find out whether they offered any advantages over traditional paper lab notebooks. The Library and the Academic Technologies group at Cornell University collaborated in carrying out a six-month pilot project of introducing this technology in research labs and science courses, using individual and classroom editions of an ELN from LabArchives. The pilot project resulted in the award of an institutional license for this ELN.

In spite of the many advantages ELNs offer compared to paper notebooks, adopting them in an environment that is using mostly paper notebooks could cause significant disruption in the research work flow (Machina and Wild, 2013a). Researchers are interested in having ELNs that are associated with laboratory instruments and scientific data management systems. Several articles have looked at some of the problems and experiences in academic laboratories that have proved challenging in implementing ELNs. They include the impact a transition to ELN may have on the scientists, the technical requirements for integration and data management, and the issues related to using third-party suppliers (Pearsall, 2013; Rudolphi and Goossen, 2011; Wright, 2009). Several articles were devoted to problems in implementing ELNs in industry (Elliott, 2013; Machina and Wild, 2013a; Metrick, 2011; Taylor, 2011).

Some of the difficulties associated with the adoption of this technology are related to data management issues, security, backup, remote access, work flow, storage (local, cloud), purchase timing, ELN integration, implementation time, potential project challenges, and external collaborators using your ELN.

The use of digital objects in research is becoming a common practice, and web-based resource models for their reuse and exchange are beginning to emerge (Lagoze et al., 2012). A study of the scholarly practices of academic chemists examined such areas as data management, research collaboration, library use, discovery, publication practices, and research funding. It also showed that chemists' needs are different from the needs of researchers in other disciplines (Bird et al., 2013; Frey and Bird, 2013).

Surveys performed at several universities about researchers' attitudes to ELNs demonstrated that the following reasons are preventing a wider adoption of this technology:

- difficulty in choosing an ELN
- uncertainty about future changes in technology
- lack of desired features
- technical problems and need of maintenance and upgrades
- effort required to transition
- intellectual property protection
- requirement analysis for an ELN

There are many questions related to ELNs that still need to be answered:

- How do ELNs and repositories interact in a particular research environment and for specific research work flows?
- How can data be entered and captured more easily?
- How can they be made available for scrutiny and reuse?
- How to add structure to data?
- Who should be involved in implementing ELNs in an institution?
- Who will decide (e.g. researcher, curator, or computer) what parameters to add to them?

As many pharmaceutical and other companies have already implemented ELNs in their work flows, exposing students to this technology will prepare them for future careers in both industry and academy and will make them more competitive. When I carried out a pilot project on ELNs in an undergraduate chemistry course at the University of Maryland College Park, the students appreciated the utility of ELNs, but they have come to realize that they were not a good fit for the course based on the way labs were organized (Baykoucheva and Friedman, 2013). Undergraduates have few data to work with, and they do not have to go back to their results. Some students prefer not to bring their laptops to labs and had first to write down the data from the experiments on paper and then reenter them in the ELN. Students concluded that the ELN did not add any value to their lab experience, except that they could familiarize themselves with this new technology. Another issue that could hinder the acceptance of an ELN in a university course is related to the fact that students do not have administrative rights, and managing their accounts could be very cumbersome and time-consuming for a single person. Another problem is the high cost of the ELNs, and there is always a question about who should be paying for the student accounts. ELNs may be more useful to graduate students, as they work with large volumes of data and often need to refer back to their previous results.

Other authors have described the implementation of an ELN in a science course. They have used an ePortfolio system, similar to an ELN. The authors concluded that the introduction of the ELN in class produced some benefits for students—they were more prepared for classes, more creative, and engaged in class. Some negative outcomes, such as increased workload for the instructor and disruption of class dynamics, were also reported (Johnston et al., 2014).

9.8 Conclusion

The new emerging ELNs are very different from earlier models. The focus now is on creating simplified and easy to use ELNs which have intuitive interfaces and can be used for calculations and data analysis. The existing proprietary models are expensive, and there is a strong interest in creating low-cost versions that do not require too much technical support. For chemists, specific features such as drawing capabilities, the ability to understand a reaction, and the accommodation of work flows appropriate to this discipline are essential. Integration with laboratory instruments is another important requirement.

Vendors are looking into expanding this technology to tablets, which combine the portability of the paper notebook with the capabilities provided by electronic solutions (Elliott, 2012). Interoperability and the capability of interfacing with scientific instruments and other laboratory management systems (LIMS) will continue to be a key requirement for most ELN implementations.

Future ELNs are expected to satisfy some of the following conditions: easy to use; platform-agnostic; enterprise-capable; supporting data publishing and archiving; affordable; and supporting intergroup collaboration. The institution should be the customer. The ELN technology should provide flexibility and breadth across disciplines, while the university will provide funding, IT support, and training. The driving forces for ELN adoption in academic institutions will be many—researchers, need to use a variety of (handheld) devices, influential groups and principal investigators, institutions represented by data librarians, and IT. Enterprise solutions, archiving and publishing capabilities, and access to data repositories of different types (e.g. domain-specific, institutional, and generic) also await development.

The transformation of a traditional lab notebook into an electronic lab notebook has not been very smooth. As ELN technology becomes more widely adopted in industrial and academic settings and becomes a reality in everyday research work flow, many researchers are trying to analyze the relationship between the technology and the emerging practices in research. The results from one study showed that there is a need for resolving "a conflict between the flexibility, fluidity, and low threshold for modifying digital records and the requirement for persistence and consistency" (Oleksik et al., 2014). Such studies could help refine the design of the new generations of ELNs.

Deciding which technology to use is very critical for the successful implementation of an ELN system. It requires a thorough analysis of all options; understanding of the culture of the organization, lab, and discipline; evaluation of users' needs, practices, research protocols, and expectations; and matching them to the products under consideration. It is also very important to take into account existing infrastructure, cost, storage options, and software maintenance. It will be a mistake to try to mold an existing ELN model after the work flow in the lab—the ELN needs to be adapted to the research environment and the needs of researchers. Those involved in making the decision should be cautious about the technology prospects and the claims made by vendors for their products.

Most of the more advanced current formats of ELNs are too expensive and not very friendly for an academic environment. Cloud technology has allowed the creation of some new, less complex and less expensive ELNs, with annual subscriptions of $100 per person, which also includes a limited cloud storage space. The ELM market is volatile and will continue to be so in the near future. Many of these ELNs are interdisciplinary, but some of them cater to chemists. The researcher communities will be driving the demand for domain-specific repositories.

Introducing ELNs in a predominantly paper-based research environment will be a difficult task. Academic libraries are strategically positioning themselves to play a role in supporting eScience. Promoting and providing training on how to use ELNs is a promising new role that subject librarians could play in their organizations.

References

Baykoucheva, S., Friedman, L., 2013. *Introducing electronic laboratory notebooks (ELNs) to students and researchers at the University of Maryland College Park.* In: Abstracts of papers of the 246th ACS National Meeting, Indiana, September 8–12, 2013.

Beato, B., Pisek, A., White, J., Grever, T., Engel, B., Pugh, M., et al., 2011. Going paperless: implementing an electronic laboratory notebook in a bioanalytical laboratory. *Bioanalysis* 3 (13), 1457–70. http://dx.doi.org/10.4155/bio.11.117.

Bird, C.L., Willoughby, C., Frey, J.G., 2013. Laboratory notebooks in the digital era: the role of ELNs in record keeping for chemistry and other sciences. *Chem. Soc. Rev.* 42 (20), 8157–75. http://dx.doi.org/10.1039/c3cs60122f.

Bogdan, K., Flowers, T., 2014. Electronic lab notebooks: supporting laboratory data in the digital era. *ISTL* (Spring). http://dx.doi.org/10.5062/F4V9861X.

Christ, C.D., Zentgraf, M., Kriegl, J.M., 2012. Mining electronic laboratory notebooks: analysis, retrosynthesis, and reaction based enumeration. *J. Chem. Inf. Model.* 52 (7), 1745–56. http://dx.doi.org/10.1021/ci300116p.

Crocker, J., Cooper, M.L., 2011. Addressing scientific fraud. *Science* 334 (6060), 1182.

Elliott, M.H., 2010. ELN in the cloud. *Sci. Comput.* 27 (2), 11–14.

Elliott, M., 2012. Tablets and ELN: a honeymoon. *Sci. Comput.* 29 (4), 9–12.

Elliott, M.H., 2013. The new frontier of biologics ELN. *Sci. Comput.* 30 (5), 4–9.

Enserink, M., 2012. Scientific ethics. Fraud-detection tool could shake up psychology. *Science* 337 (6090), 21–22.

Frey, J.G., Bird, C.L., 2011. Web-based services for drug design and discovery. *Expert Opin. Drug Discovery* 6 (9), 885–95. http://dx.doi.org/10.1517/17460441.2011.598924.

Frey, J.G., Bird, C.L., 2013. Cheminformatics and the Semantic Web: adding value with linked data and enhanced provenance. *Wiley Interdiscip. Rev. Comput. Mol. Sci.* 3 (5), 465–81. http://dx.doi.org/10.1002/wcms.1127.

He, T., 2013. Retraction of global scientific publications from 2001 to 2010. *Scientometrics* 96 (2), 555–61. http://dx.doi.org/10.1007/s11192-012-0906-3.

John, D., Banaszczyk, M., Weatherhead, G., 2011. The multidisciplinary electronic laboratory notebook: pipe dream or proven success? *Am. Lab.* 43 (11), 16–21.

Johnston, J., Kant, S., Gysbers, V., Hancock, D., Denyer, G., 2014. Using an ePortfolio system as an electronic laboratory notebook in undergraduate biochemistry and molecular biology practical classes. *Biochem. Mol. Biol. Educ.* 42 (1), 50–57. http://dx.doi.org/10.1002/bmb.20754.

Kevles, D.J., 2000. *The Baltimore Case: A Trial of Politics, Science, and Character.* W. W. Norton & Company: New York, NY.

Khaled, K.F., 2013. Scientific fraud and the power structure of science. *Res. Chem. Intermed.* http://dx.doi.org/10.1007/s11164-013-1128-x, Epub ahead of print.

Lagoze, C., Van De Sompel, H., Nelson, M., Warner, S., Sanderson, R., Johnston, P., 2012. A Web-based resource model for scholarship 2.0: object reuse & exchange. *Concurr. Comput.* 24 (18), 2221–40. http://dx.doi.org/10.1002/cpe.1594.

Machina, H.K., Wild, D.J., 2013a. Electronic laboratory notebooks progress and challenges in implementation. *J. Lab. Autom.* 18 (4), 264–8. http://dx.doi.org/10.1177/2211068213484471.

Machina, H.K., Wild, D.J., 2013b. Laboratory informatics tools integration strategies for drug discovery: integration of LIMS, ELN, CDS, and SDMS. *J. Lab. Autom.* 18 (2), 126–36. http://dx.doi.org/10.1177/2211068212454852.

MacNeil, R., 2011. The benefits of integrated systems for managing *both* samples *and* experimental data: an opportunity for labs in universities and government research institutions to lead the way. *Autom. Exp.* 3 (1), http://dx.doi.org/10.1186/1759-4499-3-2.

Maisonneuve, H., 2012. The management of errors and scientific fraud by biomedical journals: they cannot replace institutions. *Presse Med.* 41 (9 Pt 1), 853–60.

Marusic, A., 2010. Editors as gatekeepers of responsible science. *Biochem. Med.* 20 (3), 282–7. http://dx.doi.org/10.11613/BM.2010.035.

Matías-Guiu, J., García-Ramos, R., 2010. Fraud and misconduct in scientific publications. *Neurologia* 25 (1), 1–4.

Metrick, G., 2011. QA/QC: ELNs have come a long way. *Sci. Comput.* 28 (1), 19–20.

Milsted, A.J., Hale, J.R., Frey, J.G., Neylon, C., 2013. LabTrove: a lightweight, web based, laboratory "blog" as a route towards a marked up record of work in a bioscience research laboratory. *PLoS One* 8 (7), e67460. http://dx.doi.org/10.1371/journal.pone.0067460.

Noyori, R., Richmond, J.P., 2013. Ethical conduct in chemical research and publishing. *Adv. Synth. Catal.* 355 (1), 3–9. http://dx.doi.org/10.1002/adsc.201201128.

Oleksik, G., Milic Frayling, N., Jones, R., 2014. *Study of an Electronic Lab Notebook Design and Practices that Emerged in a Collaborative Scientific Environment.* In: Proceedings of 2014 ACM Conference on Computer Supported Cooperative Work and Social Computing (CSCW'14). ACM: Baltimore, MD.

Pearsall, A.B., 2013. Implementation and challenges of electronic notebooks. *Bioanalysis* 5 (13), 1609–11. http://dx.doi.org/10.4155/bio.13.75.

Rajarao, J., Weiss, S., 2011. Using E-WorkBook Suite to implement quality control in real time: expanding the role of electronic laboratory notebooks within a bioanalysis laboratory. *Bioanalysis* 3 (13), 1513–19. http://dx.doi.org/10.4155/bio.11.144.

Roland, M.-C., 2007. Publish *and* perish. Hedging and fraud in scientific discourse. *EMBO Rep.* 8 (5), 424–8. http://dx.doi.org/10.1038/sj.embor.7400964.

Rubacha, M., Rattan, A.K., Hosselet, S.C., 2011. A review of electronic laboratory notebooks available in the market today. *J. Lab. Autom.* 16 (1), 90–98. http://dx.doi.org/10.1016/j.jala.2009.01.002.

Rudolphi, F., Goossen, L.J., 2011. Electronic laboratory notebook: the academic point of view. *J. Chem. Inf. Model.* 52 (2), 293–301. http://dx.doi.org/10.1021/ci2003895.

Taylor, K.T., 2011. *Evolution of Electronic Laboratory Notebooks.* In S. Ekins, M.A.Z. Hupcey, A. J. Williams (Eds.), *Collaborative Computational Technologies for Biomedical Research.* John Wiley & Sons, Inc.: Hoboken, NJ; pp. 303–20.

Voegele, C., Bouchereau, B., Robinot, N., McKay, J., Damiecki, P., Alteyrac, L., 2013. A universal open-source electronic laboratory notebook. *Bioinformatics* 29 (13), 1710–12. http://dx.doi.org/10.1093/bioinformatics/btt253.

Walsh, E., Cho, I., 2013. Using Evernote as an electronic lab notebook in a translational science laboratory. *J. Lab. Autom.* 18 (3), 229–34. http://dx.doi.org/10.1177/2211068212471834.

Wright, J.M., 2009. Make it better but don't change anything. *Autom. Exp.* 1 (1), http://dx.doi.org/10.1186/1759-4499-1-5.

Zeitoun, J.D., Rouquette, S., 2012. Communication of scientific fraud. *Presse Med.* 41 (9 Pt 1), 872–7.

The complexity of chemical information: interview with Gary Wiggins

10

Gary Wiggins

Gary Wiggins was the head of the Indiana University (IU) Chemistry Library from 1976 to 2003. During the final four years of his professional career, he served as director of the Bioinformatics and Cheminformatics Programs in the IU School of Informatics, helping to create one of the first graduate programs in the United States that offer specialized training in cheminformatics. For many years, he taught courses in chemical information and science reference at IU. His textbook *Chemical Information Sources* was eventually converted to a Wikibook. Dr. Wiggins received several prestigious awards throughout his career, including the American Chemical Society Division of Chemical Information's Herman Skolnik Award and the Patterson-Crane Award of the ACS Columbus and Dayton Sections. He was also elected to the Special Libraries Association Hall of Fame. Much of his research involved the improvement of teaching information literacy to chemistry and science students and the improvement of communication among scientists.

Svetla Baykoucheva: In May 1991, you started a chemical information discussion list in Indiana University. Through the years, this forum became an institution of its own, providing a medium for exchanging information and ideas and attracting people interested in chemical information, but who approached it from different perspectives. Looking back at the dynamics of the Chemical Information Sources Discussion List, what do you think was the impact of this unique forum on the evolution of chemical information as a discipline, and how did it benefit those who engaged in this discourse?

Gary Wiggins: In this era of social media, it is surprising to me that an e-mail Listserv based on technology developed over 20 years ago is still thriving. In many ways, CHMINF-L is still *the* information source for everyone from chemists to chemistry/science librarians and publishers. I think that the diversity of the audience was likely the single thing that made it so valuable. Chemists were able to learn of new information sources and the tricks of the trade from librarians. Librarians were able to find help from chemists and other colleagues when reference questions were too complicated for them to understand the underlying subtleties that kept them from finding the right answers. The publishers of scientific journals and databases could avail themselves of a free resource where they could assess information needs and do market research. In that sense, CHMINF-L served to identify areas where there were gaps in chemical information and knowledge and helped to focus publishers on tools that were needed by the chemical information community. The benefit to those who participated in the discourse can be judged by the continuing support of the existing subscribers (over 1300), quite a number of whom have been on the list since its inception.

SB: The American Chemical Society (ACS) created the Division of Chemical Information (CINF) 70 years ago; this Division has been publishing a journal, the Chemical Information Bulletin, *for 65 years; Indiana University has a graduate program in chemical information; a Listserv focused on chemical information has survived for 23 years. What is so special about chemical information that made all this happen?*

GW: I would have to say that it is not the chemical information per se that makes it special, but the chemists' desire to codify and make sense of their enormously complex scientific discipline. Chemical research sometimes leads to bewildering results and often to a mountain of data that needs to be interpreted and compared to previous results. Even in the early days of modern chemistry in the nineteenth century, chemists were devising ways to communicate and archive their results for future retrieval. Although other scientific disciplines have made similar efforts, it has long been recognized that chemists led the way in organizing their literature. I have sometimes used the analogy of Mendeleev's achievement in devising the periodic table to explain the impetus for cheminformatics. When faced with a large amount of data that was confounding, Mendeleev found a way to organize and mine it to give a logical and useful tool that even made predictions about elements that hadn't yet been discovered. Likewise today, with the data deluge that modern chemistry produces, the task of cheminformaticians and other chemistry knowledge workers is to make sense out of data and organize it in ways that allow future retrieval and analysis.

SB: Is the way chemists perform research, seek information, and report their findings different from what researchers in other disciplines do?

GW: By and large, the answer to this question is no. Chemists present preliminary results at conferences and publish the final results in scientific journals, as do scientists in other disciplines. However, the web has allowed scientists to bypass the traditional sequence of publishing scientific results, most noticeably in physics, where there is a tradition of posting preprints of papers before formal publication in a refereed scientific journal. A preprint archive in chemistry was tried, but it quickly failed because there was no tradition of exchanging preprints in chemistry. There is one aspect of chemistry that makes the information-seeking activities of chemists quite unique. That is the use of the chemical structure as a universally recognized search key. Much of the organization of chemical information is predicated on the similarities of properties among chemical substances with similar structures. It was only natural that ways would be devised to search the primary chemical literature and secondary databases by structure. The culture of a scientific discipline is very hard to change. Nevertheless, when there is an overriding need that must be met, such as the need to clearly depict and search by a chemical structure, new developments will be adopted quickly by chemists. Thus, it can be predicted that the InChI (IUPAC International Chemical Identifier) will be accepted by chemists because it offers a nonproprietary identifier for chemical substances that can be used in printed and electronic data sources, allowing the linking of diverse data sources through the InChI.

SB: Indiana University has a group working on electronic laboratory notebooks (ELNs) and their implementation in industrial and academic labs. As we know, the adoption of this technology has been slow in academic institutions. What obstacles exist (technological and psychological) for the acceptance of ELNs by researchers?

GW: When I first came to Indiana University, I was responsible for a unit that produced microfiches of laboratory notebooks. It was never an idea that caught on in the Chemistry Department, but it did indicate that there was some consideration to getting the data into a format that was more easily portable and could be reproduced easily. One of our current PhD students, who is also employed in industry, has taken on some research in electronic laboratory notebooks. ELNs are an easier sell in industry because so much of the research in their labs is interrelated and driven by common goals of the company. Once a package is chosen, everyone uses it. In the academic setting, there are many research fiefdoms, and the research conducted in one group might have little or no relevance for other groups at that university. Hence, using the argument that ELNs must be used "for the good of the cause" is not very compelling in academia. On the other hand, all academic research groups have strong contacts with colleagues doing research at other academic institutions and national laboratories. It seems to me that there should be some incentive for closely related groups at multiple institutions to settle on a standard commercial ELN package that they would support. After all, it is the usual case that we see a lot of migration among such institutions by undergraduates who go on to graduate school, graduate students who accept postdoctoral fellowships at the related research groups, and faculty who move to new positions. It would make sense that one of the skills they take along with them is proficiency in the ELN software package of choice.

SB: With research becoming more and more interdisciplinary and data-intensive, how do you see the ability of academic libraries, with their limited resources, to support eScience/eResearch in their institutions? Are the libraries promising something that they won't be able to deliver? And if someone needs to provide support for eScience, would the libraries be the suitable candidate to play this role?

GW: Someone must support eScience/eResearch in academic institutions. However, a certain naivety about the data deluge can delude higher administrators in academic libraries into thinking that this is just another indexing and cataloging job, and we have been doing that for centuries. Designing a system for the effective retrieval *and* use of Big Data requires a tremendous understanding of the science and the likely uses to which the data will be put. With cloud computing becoming the norm in some areas of academic research, it is essential to understand the architecture of the current computing environment and to design effective and secure interfaces to those systems. Having said all of that, you might assume that I think academic librarians are not up to the task of supporting eScience/eResearch, but that is not the case. In fact, I think that their traditions of service to scientific faculty are valuable foundations on which to build systems that are capable of dealing with the problem. However, it is not something that can be grafted onto the existing duties of librarians by appointing an associate dean for eScience/eResearch and charging that person to layer even more work onto already overworked librarians. If such efforts are going to be based in academic libraries, the administrators—both library and university administrators—must understand that the task will require much additional training and much more staff.

SB: There are many discussions about the future of academic libraries, in general, and the new role that subject/liaison librarians will play in particular. Such new roles depend on reskilling of existing library staff and preparing a new generation of librarians trained to support research. In your opinion, are library schools in the country prepared to train librarians for such new roles?

GW: Having worked as a professional academic librarian for nearly 35 years, I have thought a lot about whether the old skill sets still have value. What I have observed is that there is always an assumption by library deans that new duties will be added, but there is rarely a corresponding assumption that some of the old responsibilities will be taken away. To the extent that library schools have a good grasp of the technological advances being made to assist researchers and give the students at least an awareness of the full range of possibilities, including the most valuable of the old skill sets, they can provide an excellent training ground for new librarians. With MOOC [massive open online course] possibilities nowadays, they can even serve the retraining needs for librarians who are at various stages of their careers. The trend toward iSchools has been unrelenting in the last several decades. Indiana University was one of the last remaining bastions where a library school maintained a separate identity from the more computer-intensive informatics/computing science departments or schools. Now, the School of Library and Information Science at IU has been absorbed by the School of Informatics and Computing. I have long supported such a union of the two schools, and I anticipate a synergy that will lead to better preparation of practitioners along the whole spectrum from computer scientists and informaticians to information scientists and librarians.

SB: Information literacy is a central focus for academic librarians. With the technological advancements and the dramatic changes in users' information needs and expectations, what role (if any) could librarians play in this new learning environment and what obstacles might they encounter in trying to adjust to these new developments?

GW: I was glad to see information literacy come into its own as a defined specialization in academic libraries in recent years. Whatever technological changes may occur in the future, a fundamental approach to information acquisition and evaluation that goes far beyond googling should always be taught. My old mentor Herb White used to say about scientists doing their own online searches, "Yes, they do it, but they do it badly!" Today's end users have bypassed the command-driven approaches to searching online databases and are familiar only with user-friendly front-end systems such as SciFinder. Nevertheless, there are many intricacies of the Chemical Abstracts and MEDLINE databases that are masked by such a system. The single most important thing that people who are charged with teaching information literacy can do is to make budding scientists aware of the rich range of information tools available to them and open their eyes to the advantages of choosing those tools over Internet search engines for almost all questions or information needs they might have. Certainly, librarians are most well suited to play the key role in information literacy on campus. The main obstacle they face is the faculty members who believe that the approach to information they have always used is the one that they should transmit to their students. It takes a gentle sales job to convince faculty members to try another approach and if not delegate the information literacy instruction to librarians, then at least share it with them. Although I have been out of the profession for some years now, it is safe to conjecture that there are still economic barriers to providing adequate information literacy training in some academic institutions. Librarians are best suited to negotiate with database publishers reasonable fees for the educational use of their products.

SB: *Some of the people who started the Open Access Movement—Vitek Tracz, for example—have moved on to other ventures such as Faculty of 1000 (http://f1000. com) and F1000Research (http://f1000research.com), which try to address issues in the "post-open access" world. There are predictions that the journal will disappear and be replaced by individual articles. In your opinion, how will the STEM publishing field change in the next few years?*

GW: I have always viewed the Faculty of 1000 as a special type of review serial. Any librarian who has ever had to conduct a serials cancellation project knows how fiercely the faculty will fight for the retention of review serials. I sometimes found that it was easier to cancel even an American Chemical Society journal than to remove an *Annual Review of* [fill in the blank] from our subscription list. The volume of primary scientific journal articles is immense, and scientists must have a mechanism that lets them focus on the articles most worthy of their attention. For years, they have depended on the reputation of a journal publisher, the editorial board of the journal, and the thoroughness with which the articles are subjected to peer review as key filters in their selection of the literature. Various alerting services have been developed to make sure they do not overlook a critical article, but even with those, key articles are sometimes missed. If we are to eliminate the journal as the unit of publication, then there must be something that substitutes for the confidence that a scientist has in the reputation of the journal itself. The move to electronic journals has opened the door to consideration of a system that would jettison the journal itself in favor of the article as the unit of publication. That may happen, but I do not foresee it coming to pass until there is a different architecture in place that gives the individual scientist a large measure of assurance that what is being read is not in fact junk science. The Faculty of 1000 is a novel way of assessing scientific journal output, but ultimately, there will be in place a mechanism that will assess the data itself behind a scientific article and give a rating to the scientist about both the validity of the research paper and the relevance of the research itself to a scientist's own research. There was at one time an attempt to characterize compilations of numeric data as being recommended (of the highest quality), provisional, typical, or selected. Data were assigned to these categories on the basis of checks against theoretical values, experimental detail, etc. What should ultimately emerge is a system using the power of cloud computing that allows such evaluation of data and articles to take place automatically and to be tied to an alerting service when items of potential interest are identified.

Publications by Gary Wiggins:

Lee, W.M., Wiggins, G., 1997. Alternative methods for teaching chemical information to undergraduates. *Sci. Technol. Libr.* 16(3/4), 31–43.

Wiggins, G., 1995. CHMINF-L, the chemical information sources discussion list. *J. Am. Soc. Inf. Sci.* 46(8), 614–17.

Wiggins, G., 1998. Chemistry on the Internet: the library on your computer. *J. Chem. Inf. Comput. Sci.* 38(6), 956–65.

Wild, D.J., Wiggins, G.D., 2006. Challenges for chemoinformatics education in drug discovery. *Drug Discov. Today* 11(9–10), 436–9.

Wild, D.J., Wiggins, G.D., 2006. Videoconferencing and other distance education techniques in chemoinformatics teaching and research at Indiana University. *J. Chem. Inf. Model.* 46(2), 495–502.

Measuring academic impact

11

All metrics are wrong, but some are useful.

(Kraker, 2014)

The impact factor is a very useful tool for evaluation of journals, but it must be used discretely. Considerations include the amount of review or other types of material published in a journal, variations between disciplines, and item-by-item impact.

Eugene Garfield

11.1   Introduction

When Eugene Garfield created the Science Citation Index (SCI) and the journal impact factor (IF) in the early 1960s, he could not have imagined that they would be used for purposes for which they had never been designed nor could he foresee the dramatic consequences his brilliant idea would have on science and the life of scientists in decades to come. For more than 50 years, the IF has been used to evaluate scientific contributions, becoming an obsession for researchers, editors of scientific journals, and administrators in academic organizations. As Eugene Garfield points out in his interview, included in Chapter 12 of this book, the IF was designed to evaluate journals, but it has been used incorrectly to evaluate authors' contributions. The SCI has provided the foundations for the Web of Science (WoS), one of the most widely used scientific databases today, and such tools for evaluating research impact as *Journal Citation Reports* (*JCR*) and Essential Science Indicators (ESI).

11.2   The Institute for Scientific Information (ISI)

The Institute for Scientific Information (ISI) (now part of Thomson Reuters) was founded by Eugene Garfield in 1960 in Philadelphia. It produced the only available bibliographic databases from which large-scale bibliometric indicators could be compiled. ISI also published the *Current Contents*, a weekly journal, with separate editions for major disciplines, that included the table of contents of selected scientific journals (Garfield, 1979). In almost each issue of the journal, there was an essay by Dr. Garfield devoted to a wide range of topics (Garfield, 2015). A chapter on the history of ISI by Bonnie Lawlor (Lawlor, 2014) was published in the book *The Future of the History of Chemical Information* (McEwen and Buntrock, 2014).

11.3   The *Science Citation Index* (*SCI*)

ISI's citation indexes, which are now included in the WoS, were the most important sources of bibliometric information until Scopus was launched by Reed Elsevier in 2004. Created by Eugene Garfield in the early 1960s, the *SCI* evolved to become the basis of innovative concepts and products such as the WoS, *JCR*, and ESI (Garfield, 1964, 1972).

The *SCI* collects citations in articles published in scientific journals, selected on the basis of their quality. To evaluate the quality of journals and help in selecting journals to be covered by the *SCI*, Garfield introduced the Journal Impact Factor (IF) (Garfield, 1972). At the beginning, the *SCI* covered 800 journals, but today, their number is more than 10,000. One of the criticisms directed at the *SCI* is that it covers mostly English-language journals. The references in the *SCI* are arranged to show how many times each publication has been cited within a certain period of time and by whom, and the results are published as the *SCI*.

From the *SCI*, new research areas such as webometrics, infometrics, and sciento-metrics emerged (Baynes, 2012).

11.4   Journal Impact Factor (IF)

On the basis of the *SCI* and authors' publication lists, the annual citation rate of papers by an author or research group can be calculated. Similarly, the citation rate of a scientific journal—known as the IF—can be calculated as the mean citation rate of all the articles contained in the journal (Garfield, 1972, 1997, 2001, 2006, 2014). There are many reasons why the *SCI* and the IF have attracted so much attention:

- *Authors* look at the IFs when deciding where to publish their articles, because they are often evaluated, hired, promoted, and funded on the basis of whether they have published in high-impact journals.
- *Funding agencies* are placing an increased weight on the IFs of the journals in which the applicants for grants have published their papers.
- *Editors* are trying to understand how the IF is calculated so that they can manipulate the content of their journals to increase their ranking. Publishers and editors can determine the journals' influence in the marketplace and review editorial functions.
- *Librarians* often make decisions about which journals to drop or add on the basis of their IFs.
- *Administrators* monitor bibliometric and citation patterns to make strategic and funding decisions.

11.5   *Journal Citation Reports* (*JCR*)

Many people use the term IF, but very few understand what lies behind it. The IF is a ratio of the number of times ***all*** items in a journal were cited, divided by the number of ***citable articles*** published in this journal for the same period of time. For example, to calculate the IF of a journal for the year 2013, Thomson Reuters, the current publisher of *JCR*, counts the number of citations made in 2013 of papers published in this journal in the previous two years, 2011 and 2012, and divides this number by the number of "citable articles" published during that two year period.

IFs, which are published annually in *SCI JCR*, are widely regarded as a measure of the quality of journals. *JCR* analyzes and summarizes citations from science and social science journals, as well as from proceedings that are collected in the WoS database. When applying for tenure, promotion, and funding, authors can find in *JCR* the performance of the journals in which they have published their articles.

The *JCR* published in a particular year reports data from the previous year. For example, the issue of the *JCR* published in 2014 reports the IFs for the year 2013 and is called *Journal Citation Reports 2013 Science Edition* (Figure 11.1). A review of Garfield's IF provides more information on how the IF is calculated and implemented (Vanclay, 2012).

11.6   The Journal Impact Factor is not without drawbacks

Although there is general agreement about the importance of publications in refereed journals with high IF as a performance indicator, more and more voices are expressing doubts about the way citations and IFs are being used to evaluate the performance of researchers and organizations and the quality of scientific journals (Brody, 2013; Delgado-Lopez-Cozar and Cabezas-Clavijo, 2013; Elkins et al., 2010; Elsaie and Kammer, 2009; Falagas et al., 2008; Glanzel and Moed, 2013; Hodge and Lacasse, 2011; Moed and Van Leeuwen, 1995; Nature Neuroscience, 1998; Zucker, 2013).

The main controversy is related to the way IFs are calculated and what constitutes a "citable" article. When the IF is calculated, the numerator is the total number of citations to any item in the journal, whereas the denominator includes only the number of articles published in the journal, excluding editorials, letters, and comments.

Some of the main drawbacks of the IF are listed below:

- concealed and flawed calculation
- does not take into account self-citations or negative citations
- includes a limited number of journals and is English-language-biased
- can be manipulated by editors (e.g. by publishing reviews as editorials and requesting that authors submitting papers cite the journal as a condition for accepting their paper).
- authors may avoid citing their competitors.
- some research fields are smaller, and a smaller pool of people could cite papers in journals in this subject area.

11.7   *Essential Science Indicators (ESI)*

Another more recent derivative of the *SCI* is *ESI*, which analyzes researchers' performance and identifies emerging research areas (Figure 11.2). It shows "who the most influential individuals, institutions, papers, publications, and countries" are in a particular field of research (Thomson Reuters, 2014).

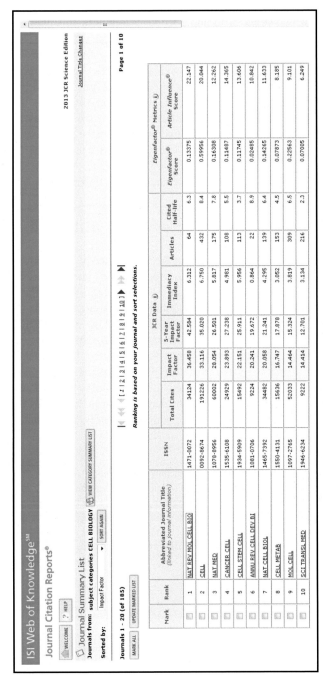

Figure 11.1 *Journal Citation Reports (JCR).* With permission from Thomson Reuters.

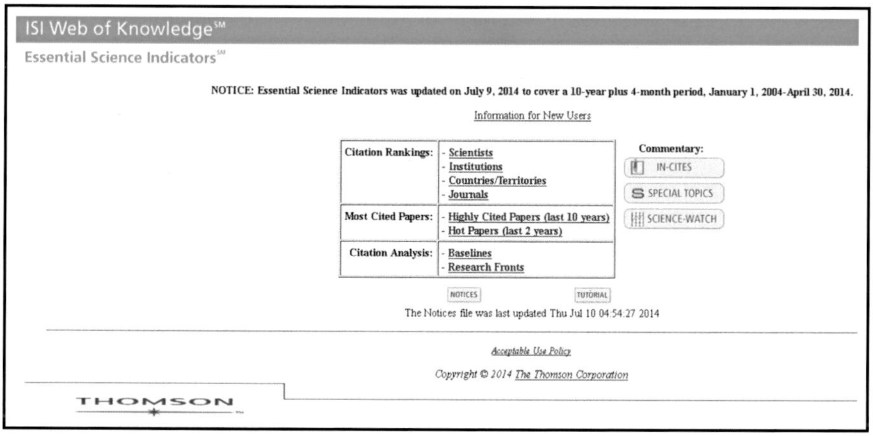

Figure 11.2 Essential Science Indicators. With permission from Thomson Reuters.

ESI can be used to analyze the scientific literature to view the research performance of companies, institutions, nations, and journals. It tracks significant trends in the sciences and social sciences; ranks top countries, journals, scientists, papers, and institutions by field of research; shows research output and impact in specific fields of research; and allows the evaluation of potential employees, collaborators, reviewers, and peers.

11.8 h-Index

The h-index is another method for evaluating researchers' accomplishments. It was developed by J.E. Hirsch in 2005 (Hodge and Lacasse, 2011; Jacso, 2012b; Norris and Oppenheim, 2010). The value of h is equal to the number of papers (N) in the list that have N or more citations. An h-index of 20 means there are 20 items that have 20 citations or more. This metric is useful because it discounts the disproportionate weight of highly cited papers or papers that have not yet been cited. It favors academics who publish a continuous stream of papers with lasting and above-average impact.

11.9 Google Scholar Citations

Google Scholar (GS) Citations allows authors to keep track of citations of their articles and check who is citing their publications. Citations in GS can be graphed over time, and several citation metrics can be calculated and updated automatically. If an author's profile is made public, it will appear in the GS results when people search for this author's name. Authors can choose to have their lists of articles updated automatically or do it manually.

Many people are wondering why the citation counts for individual authors are higher in GS than in WoS. Many recent publications have compared this tool with citations provided by other sources (Aguillo, 2012; Delgado-Lopez-Cozar and Cabezas-Clavijo, 2013; Gaze, 2014; Hightower and Caldwell, 2010; Jacso, 2005, 2008, 2012a; Kulkarni et al., 2009; Parry, 2014; Shultz, 2007; Snyder, 2012).

A detailed analysis of the citations found in GS and WoS for a randomly selected paper showed that GS is a better predictor of actual citation numbers than is WoS (Gaze, 2014). Another author compared citation counts from GS, Scopus, and WoS and found that GS showed a significantly higher number of citations than Scopus or WoS. And this higher number did, indeed, include a patent, books, dissertations, a preprint of an article, and even a grant application, but even subtracting these, the author found that GS caught more citations than the other two databases (Snyder, 2012).

11.10   How do authors decide what and how to cite?

New science builds on previous achievements, and it is important to cite properly what was done in the past. Many years ago, I got myself embroiled in an international intrigue. While browsing through some new issues of journals in the library of my institute, I came across a review article in which the author has cited a very important, pivotal paper, not in the reference list, but in the text of the article. I do not know what pushed me to make a comment about this unusual method of citation, in a letter to one of the authors of the cited article. What I did not expect to happen, though, was that a copy of my letter (without showing my name) would be sent to the author of the review about which I had made this comment.

Later, I was shown the correspondence about my comment between the author of the review and the author of the cited article and was surprised to see how the author of the review had explained why he decided to cite the article in the text. The reason, he said, was a limitation of the number of references that the journal had allowed per article. No matter what one may think of this incident, it shows at least two things—that scientists do care how they will be perceived by others and especially by their competitors. It also exemplified how judicious the selection and use of citations might be. It has become more common for authors to cite a review paper, rather than original communications, which is another factor impacting the citation record of a researcher.

So, how do researchers decide what to cite in their publications? And what are their biases? Is there a tit-for-tat game—you cite my papers and I will cite yours? In an interview with Eugene Garfield (Baykoucheva, 2006), I asked him about some practices of researchers in selecting what to cite.

> *Svetla Baykoucheva: Do such factors as the pressure to get cited influence an author's decisions which papers to cite? For example, are there incidents when friends cite each other's papers, even if there are more relevant papers to cite? Or former graduate students and post-docs continue to cite their former mentors indefinitely?*

> Eugene Garfield: There are a lot of anecdotal claims on these issues, but little systematic evidence. If you stay in the same field, it is inevitable and justifiable that you will cite certain mentors "indefinitely." The relevance of Robert K Merton's work in my writings didn't change over the 40 years I knew him; so I continued to cite his work and that of many others whenever it was appropriate. When people write brief introductory historical statements for new papers they will cite one or more of the same people who started the field. That may be why co-citation clustering works.

Nationality bias in selecting certain articles over others to cite was discussed in Chapter 3.

11.11   More on evaluating journals

11.11.1   Journal metrics from Elsevier

Since scientists are judged by the quality of the journals in which they have published, choosing the right journals could be of critical importance for researchers' careers, promotions, and research funding. The IF should not be used alone, though, in assessing the usefulness of a journal. Elsevier provides free tools to track, analyze, and visualize research output, using three different impact metrics based on methodologies that use Scopus as the data source (Figures 11.3–11.6). The figures are reproduced with permission from Scopus and Elsevier.

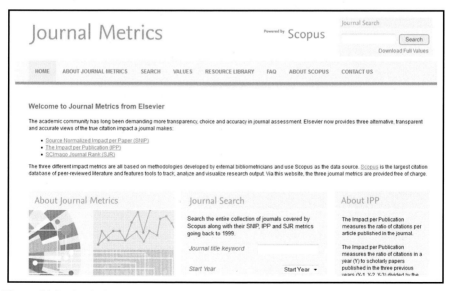

Figure 11.3 Elsevier's Journal Metrics page is an access point for using three different metrics to evaluate journals.

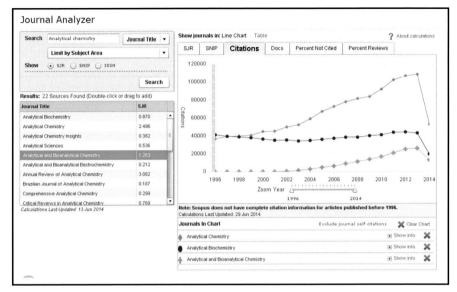

Figure 11.4 The Scopus Journal Analyzer shows citation data for three chemistry journals, from 1996 to 2014.

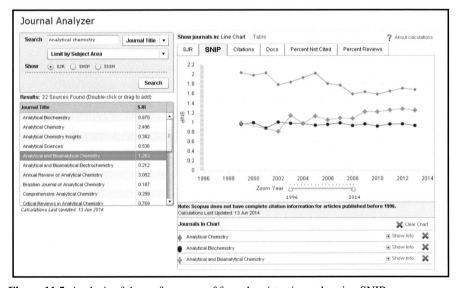

Figure 11.5 Analysis of the performance of four chemistry journals using SNIP.

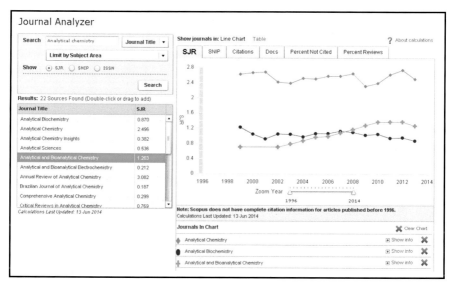

Figure 11.6 SCImago Journal Rank (SJR) of three chemistry journals.

11.11.2 Source Normalized Impact per Paper (SNIP)

The SNIP weighs citations based on the total number of citations in a subject field to calculate the contextual citation impact. This is especially helpful to researchers publishing in multidisciplinary fields, as it normalizes different subject areas, allowing researchers, authors, and librarians to compare journals from such different subject areas.

11.11.3 The Impact per Publication (IPP)

The IPP metric is using a citation window of three years, which is considered to be the optimal time period to accurately measure citations in most subject fields. It measures the ratio of citations in a year (Y) to scholarly papers published in the three previous years (Y-1, Y-2, and Y-3), divided by the number of scholarly papers published in those same years (Y-1, Y-2, and Y-3). The IPP takes into account the same peer-reviewed scholarly papers in both the numerator and denominator of the equation to provide a fair-impact measurement of the journal, which diminishes the chance of manipulation.

11.11.4 SCImago Journal Rank (SJR)

SJR is based on the idea that "all citations are not created equal." It is a measure of prestige of a journal, ranking journals by their "average prestige per article." The subject field, quality, and reputation of the journal directly impact the value of a citation.

11.12   Conclusion

Scientists occasionally talk to colleagues about scientific impact. This discussion often focuses on which journals have higher IFs and which papers have received many citations. Although the value of a contribution does not entirely correlate with the number of times it is cited, the number of citations a paper has received is much more important than the fact that it was published in a journal with high IF.

Each source of citation data gathers citations in its own ways, with their own strengths and limitations. WoS gets citation counts by manually gathering citations from a relatively small set of "core" journals. Scopus and GS crawl a much more expansive set of publisher web pages, and Google also examines papers hosted elsewhere on the web. PubMed looks at the reference sections of papers in PubMed Central, and CrossRef looks at the reference lists that they see. GS's gathering techniques and citation criteria are the most inclusive; the number of citations found by GS is typically the highest, though the least curated. Many authors have looked into the differences between citation counts from different providers, comparing GS, Scopus, and WoS and finding many differences (Archambault et al., 2009; Elkins et al., 2010; Falagas et al., 2008; Haustein et al., 2014; Jacso, 2008; Kulkarni et al., 2009; Torres-Salinas et al., 2009).

Whether we want to admit it or not, the *SCI* and the IF are having a tremendous impact on many aspects of science. Analyses of citations have even been used to predict who will get the next Nobel Prizes. For no better measurement in evaluating scientific research, the IF continues to be taken into account in evaluating the performance of researchers and journals (Brody, 2013; Thelwall et al., 2013). Chapter 14 discusses the emerging area of altmetrics, which uses alternative approaches to measure attention to research.

References

Aguillo, I.F., 2012. Is Google Scholar useful for bibliometrics? A webometric analysis. *Scientometrics* 91 (2), 343–351. http://dx.doi.org/10.1007/s11192-011-0582-8.

Archambault, E., Campbell, D., Gingras, Y., Larivière, V., 2009. Comparing bibliometric statistics obtained from the Web of Science and Scopus. *J. Am. Soc. Inf. Sci. Technol.* 60 (7), 1320–6. http://dx.doi.org/10.1002/asi.21062.

Baykoucheva, S., 2006. Interview with Eugene Garfield. *Chemical Information Bulletin*, 58 (2), 7–9. http://acscinf.org/content/interview-eugene-garfield.

Baynes, G., 2012. Scientometrics, bibliometrics, altmetrics: some introductory advice for the lost and bemused. *Insights* 25 (3), 311–15. http://dx.doi.org/10.1629/2048-7754.25.3.311.

Brody, S., 2013. Impact factor: imperfect but not yet replaceable. *Scientometrics* 96 (1), 255–7. http://dx.doi.org/10.1007/s11192-012-0863-x.

Delgado-López-Cózar, E., Cabezas-Clavijo, Á., 2013. Ranking journals: could Google Scholar Metrics be an alternative to Journal Citation Reports and Scimago Journal Rank? *Learn. Publ.* 26 (2), 101–13. http://dx.doi.org/10.1087/20130206.

Elkins, M.R., Maher, C.G., Herbert, R.D., Moseley, A.M., Sherrington, C., 2010. Correlation between the Journal Impact Factor and three other journal citation indices. *Scientometrics* 85 (1), 81–93. http://dx.doi.org/10.1007/s11192-010-0262-0.

Elsaie, M.L., Kammer, J., 2009. Impactitis: the impact factor myth syndrome. *Indian J. Dermatol.* 54, 83–5. http://dx.doi.org/10.4103/0019-5154.48998.

Falagas, M.E., Kouranos, V.D., Arencibia-Jorge, R., Karageorgopoulos, D.E., 2008. Comparison of SCImago journal rank indicator with journal impact factor. *FASEB J.* 22 (8), 2623–8. http://dx.doi.org/10.1096/fj.08-107938.

Garfield, E., 1964. "Science Citation Index"—new dimension in indexing. *Science* 144 (3619), 649–54. http://dx.doi.org/10.1126/science.144.3619.649. http://garfield.library.upenn.edu/essays/v7p525y1984.pdf.

Garfield, E., 1972. Citation analysis as a tool in journal evaluation. *Science* 178 (4060), 471–9.

Garfield, E., 1979. *Current Contents*: its impact on scientific communication. *Interdiscip. Sci. Rev.* 4 (4), 318–23. http://dx.doi.org/10.1179/isr.1979.4.4.318. http://garfield.library.upenn.edu/essays/v6p616y1983.pdf.

Garfield, E., 1997. All sorts of authorship. *Nature* (London, UK) 389 (6653), 777. http://dx.doi.org/10.1038/39706.

Garfield, E., 2001. Impact factors, and why they won't go away. *Nature* (London, UK) 411 (6837), 522. http://dx.doi.org/10.1038/35079156. http://garfield.library.upenn.edu/papers/nature411p522y2001.htm.

Garfield, E., 2006. The history and meaning of the Journal Impact Factor. *JAMA* 295 (1), 90–93. http://dx.doi.org/10.1001/jama.295.1.90. http://garfield.library.upenn.edu/papers/jamajif2006.pdf.

Garfield, E., 2014. *Papers on Impact Factor*, Retrieved 22 August, 2014, from http://www.garfield.library.upenn.edu/impactfactor.html.

Garfield, E., 2015. *Essays of an Information Scientist*. Retrieved May 18, 2015, from http://www.garfield.library.upenn.edu/essays.html.

Gaze, W., 2014. *Citation counts: Google Scholar vs. Web of Science*, Retrieved August 29, 2014, from http://coastalpathogens.wordpress.com/2014/01/08/citation-counts-google-scholar-vs-web-of-science.

Glanzel, W., Moed, H.F., 2013. Opinion paper: thoughts and facts on bibliometric indicators. *Scientometrics* 96 (1), 381–94. http://dx.doi.org/10.1007/s11192-012-0898-z.

Haustein, S., Peters, I., Sugimoto, C.R., Thelwall, M., Larivière, V., 2014. Tweeting biomedicine: an analysis of tweets and citations in the biomedical literature. *J. Assoc. Inf. Sci. Technol.* 65 (4), 656–69. http://dx.doi.org/10.1002/asi.23101.

Hightower, C., Caldwell, C., 2010. Shifting sands: science researchers on Google Scholar, Web of Science, and PubMed, with implications for library collections budgets. *ISTL* (Fall). http://dx.doi.org/10.5062/F4V40S4J.

Hodge, D.R., Lacasse, J.R., 2011. Evaluating journal quality: is the h-index a better measure than impact factors? *Res. Soc. Work Pract.* 21 (2), 222–30. http://dx.doi.org/10.1177/1049731510369141.

Jacsó, P., 2005. As we may search—comparison of major features of the Web of Science, Scopus, and Google Scholar citation-based and citation-enhanced databases. *Curr. Sci.* 89 (9), 1537–47.

Jacsó, P., 2008. The pros and cons of computing the h-index using Google Scholar. *Online Inf. Rev.* 32 (3), 437–52. http://dx.doi.org/10.1108/14694520810889718.

Jacsó, P., 2012a. Grim tales about the impact factor and the h-index in the Web of Science and the Journal Citation Reports databases: reflections on Vanclay's criticism. *Scientometrics* 92 (2), 325–54. http://dx.doi.org/10.1007/s11192-012-0769-7.

Jacsó, P., 2012b. Using Google Scholar for journal impact factors and the h-index in nationwide publishing assessments in academia—siren songs and air-raid sirens. *Online Inf. Rev.* 36 (3), 462–78. http://dx.doi.org/10.1108/14684521211241503.

Kraker, P., 2014. *All metrics are wrong, but some are useful*. Retrieved June 26, 2014, from http://science.okfn.org/tag/altmetrics/.

Kulkarni, A.V., Aziz, B., Shams, I., Busse, J.W., 2009. Comparisons of citations in Web of Science, Scopus, and Google Scholar for articles published in general medical journals. *JAMA* 302 (10), 1092–6.

Lawlor, B., 2014. The Institute for Scientific Information: A Brief History. In: McEwan, L.R., Buntrock, R.E. (Eds.), *The Future of the History of Chemical Information, ACS Symposium Series*. American Chemical Society: New York, NY, pp. 109–26.

McEwen, L.R., Buntrock, R.E. (Eds.), 2014. *The Future of the History of Chemical Information*. American Chemical Society: New York, NY.

Moed, H.F., Van Leeuwen, T.N., 1995. Improving the accuracy of Institute for Scientific Information's journal impact factors. *J. Am. Soc. Inf. Sci.* 46 (6), 461–7. http://dx.doi. org/10.1002/(SICI)1097-4571(199507)46:6<461::AID-ASI5>3.0.CO;2-G.

Moed, H.F., van Leeuwen, T.N., 1996. Impact factors can mislead. *Nature* 381 (6579), 186.

Nature Neuroscience, 1998. (Editorial) Citation data: the wrong impact? *Nat. Neurosci.* 1 (8), 641–2.

Norris, M., Oppenheim, C., 2010. The h-index: a broad review of a new bibliometric indicator. *J. Doc.* 66 (5), 681–705. http://dx.doi.org/10.1108/00220411011066790.

Parry, M., 2014. As researchers turn to Google, libraries navigate the messy world of discovery tools. *Chron. High. Educ.*, April 21. Retrieved from http://chronicle.com/article/As-Researchers-Turn-to-Google/146081/

Shultz, M., 2007. Comparing test searches in PubMed and Google Scholar. *J. Med. Libr. Assoc.* 95 (4), 442–5.

Snyder, J., 2012. *Google Scholar vs. Scopus & Web of Science*. Retrieved from http://www.functionalneurogenesis.com/blog/2012/02/google-scholar-vs-scopus-web-of-science.

Thelwall, M., Haustein, S., Larivière, V., Sugimoto, C.R., 2013. Do altmetrics work? Twitter and ten other social web services. *PLoS One* 8 (5), 7. http://dx.doi.org/10.1371/journal.pone.0064841.

Thomson Reuters, 2014. *Essential Science Indicators*, Retrieved 22 August, 2014, from http://thomsonreuters.com/essential-science-indicators/.

Torres-Salinas, D., Ruiz-Perez, R., Delgado-Lopez-Cozar, E., 2009. Google Scholar as a tool for research assessment. *Prof. Inf.* 18 (5), 501–10. http://dx.doi.org/10.3145/epi.2009.sep.03.

Vanclay, J.K., 2012. Impact factor: outdated artefact or stepping-stone to journal certification? *Scientometrics* 92 (2), 211–38.

Zucker, K.J., 2013. The Impact Factor: just the facts. *Arch. Sex. Behav.* 42 (4), 511–14. http://dx.doi.org/10.1007/s10508-013-0105-1.

From the Science Citation Index to the Journal Impact Factor and Web of Science: interview with Eugene Garfield

Eugene Garfield

I had always visualized a time when scholars would become citation conscious and to a large extent they have, for information retrieval, evaluation and measuring impact… I did not imagine that the worldwide scholarly enterprise would grow to its present size or that bibliometrics would become so widespread.

Eugene Garfield

The most remarkable contribution of Eugene Garfield to science is the creation of the *Science Citation Index (SCI)*. In the early 1960s, he founded in Philadelphia the Institute for Scientific Information (ISI) (now part of Thomson Reuters), which became a hotbed for developing new information products (Figure 12.1). Based on the concept of using articles cited in scientific papers to categorize, retrieve, and track scientific and scholarly information, the *SCI* was further developed to create the Web of Science, one of the most widely used databases for scientific literature today. *Journal Citation Reports*, which publishes the impact factors of journals, and *Essential Science Indicators*, which provides information about the most influential individuals, institutions, papers, and publications, are also based on the *SCI*. The concept of "citation indexing," propelled by the *SCI*, triggered the development of new fields such as bibliometrics, informetrics, and scientometrics.

Managing Scientific Information and Research Data

Figure 12.1 The Institute for Scientific Information at 3501 Market Street, Philadelphia, now part of Drexel University.

Eugene Garfield's career is marked also by the development of other innovative information products that include *Index Chemicus, Current Chemical Reactions*, and *Current Contents*. The latter included the tables of contents of many scientific journals and had editions covering clinical medicine, chemistry, physics, and other disciplines. Citation indexes for the *Social Sciences Citation Index (SSCI)* and the *Arts & Humanities Citation Index (A&HCI), Index to Scientific & Technical Proceedings and Books (ISTP&B)*, and *Index to Scientific Reviews* were later added to this list.

Garfield started the *SCI* and *Current Contents (CC)* in a chicken coop in New Jersey (shown in Figure 12.2). For almost every issue of *CC* he wrote an essay that

Figure 12.2 Dr. Garfield in front of the chicken coop in New Jersey where *Current Contents* and the *Science Citation Index* were created.

Figure 12.3 Pictures of Eugene Garfield published with his essays in editions of *Current Contents.*

was accompanied by his picture (Figure 12.3). These essays, devoted to a broad range of topics, have fascinated scientists and information specialists from all over the world.

The *SCI* preceded the search engines, which utilized the principle of citation indexing to create algorithms for relevancy of documents. "Citation linking," a concept that is central to the *SCI*, was on Sergey Brin's and Larry Page's minds when they published the paper in which Google was first mentioned (Brin and Page, 1998).

The *SCI* has been used in ways that its creator never envisioned. The Journal Impact Factor (IF), derived from the *SCI* and created for the purpose of evaluating journals, was (incorrectly) extrapolated to measure the quality of research of individual scientists. IF has been used by information scientists, research administrators, and policy makers to see trends in scientific communication and to compare countries, institutions, departments, research teams, and journals by their productivity and impact in various fields. Sociologists and historians of science have been studying processes, phenomena, and developments in research using data from the *SCI*. Librarians often use the IF to select and "weed out" their collections. Editors monitor their journal's impact and citation, and publishers use it to market their publications and decide whether to launch new journals or discontinue existing ones.

It is not an exaggeration to say that the creation of the *SCI* was one of the most significant events in modern science, in general, and in scientific information, in particular. The celebration of the golden anniversary of the *SCI* in May 2014 was also a celebration of the incredible legacy of Eugene Garfield.

Svetla Baykoucheva: At the time when you created the Science Citation Index (SCI) *(Garfield, 1964), you could not have imagined that it would have such ramifications for science. How did you come up with the idea of using citations in articles to retrieve, organize, and manage scientific information and make it discoverable?*

Eugene Garfield: When I was working at the Welch Medical Library at Johns Hopkins University in 1951–1953, I was advised by Chauncey D. Leake to study the role and importance of review articles and decided to study their linguistic structure. I came to the conclusion that each sentence in a review article is an indexing statement. So I was looking for a way to structure them. Then I got a letter from W. C. Adair, a retired vice president of *Shepard's Citations*. He told me about correspondence he had back in the 1920s with some scientists about the idea of creating an equivalent to *Shepard's Citations* for science. *Shepard's Citations* is a legal index to case law. It uses a coding system to classify subsequent court decisions based on earlier court cases. If you know the particular case, you can find out what decisions, based on that case, have been appealed, modified, or overruled.

In *Shepard's*, these commentaries were organized by the cited case, that is, by citation rather than by keywords as in traditional science indexes. I began to correspond with Mr. Adair. I was then a new associate editor of *American Documentation*. So I suggested that he wrote an article for the journal, and he did. Then, I wrote a follow-up article in *Science* in 1955 on what it would mean to have a similar index for science literature.

While *Shepard's* dealt with thousands of court cases, the scientific literature involves millions of papers; so for practical reasons, an index for the sciences was orders of magnitude greater. Since I could not get any government support, I left the indexing project. In 1953, I went back to Columbia University to get my master's degree in library science. I was told by NSF that I could not apply for a grant, because I was not affiliated with any educational or nonprofit institution.

After I wrote the article in *Science*, I got a letter from Joshua Lederberg who won the Nobel Prize in 1958. He said that for lack of a citation index, he could not determine what had happened after my article was published. He then told me that there was an NIH study section of biologists and geneticists. He suggested that I wrote a proposal for an NIH grant. NIH gave ISI a grant of $50,000 a year for three years. However, after the first year, when Congressman Fountain of North Carolina questioned NIH policy, they decided that grants could not be made to for-profit companies. So NIH transferred the funds to the NSF and that's how it became an NSF-funded project.

We eventually published the *Genetic Citation Index*, which is described in the literature. We printed 1000 copies and distributed them to individual geneticists. We produced the 1961 *Science Citation Index* as part of the *GCI* experiment. It covered 613 leading journals. We asked NSF to publish it, but they refused. At that time, we were very successful with *Current Contents*, so ISI used company funds to finance the launch of the 1961 *Science Citation Index* and began regular quarterly publication in 1964.

SB: You have founded the Institute for Scientific Information in Philadelphia, which has become the most important center for scientific information in the world, but many people who now use Web of Science don't even realize that it is based on the SCI. *How did the SCI "underwrite" such resources as the Web of Science,* Journal Citation Reports, *and* Essential Science Indicators?

EG: When we first started the *SCI*, it was a print product. From the very beginning, we used punched cards. So, in a sense, it was machine-readable. Later, we also published the *SCI* on CD-ROM and we released our magnetic tapes to various institutions, including the NSF. The CD-ROM version evolved, and then in 1972, *SCI* went online. The ISI databases *SCI* and *SSCI* were among the first to be on DIALOG. *SSCI* was database #6. That's how ISI databases evolved into the electronic form known today as *Web of Science*. Each year, we increased the source journal coverage. We started with 615 source journals in 1961. I think that more than 10,000 are now covered.

This is how the *SCI* works: whatever you cite in your published papers will be indexed in the database. This is true whether or not the cited journal or book is covered as a source in the database. However, even if an article is covered as a source item, but is never cited, it will not show up in the "Cited Reference" section. You need to differentiate between the source index and the citation index. You use the source index to find out what has been published in the journals that we cover and you use the citation index to see what has been cited. Web of Science includes both the source index and the citation index.

SB: One by-product of the SCI *was the Journal Impact Factor, which has become a major system for evaluating scientific impact. You have been quoted saying that the impact factor is an important tool, but it should be used with caution. When should the IF not be used?*

EG: It is not appropriate to compare articles by IF, because IF applies to an entire journal. While the IF of the journal may be high, an individual article published in it may never be cited. You publish a paper, for example, in *Nature*, which has very high IF. The article may never be cited, but if it is published in a high-impact journal, it indicates a high level of quality by being accepted in that journal. I have often said that this is not a proper use of the IF. It is the citation count for an individual article and not the IF of the journal, which matters most. The Journal IF is an average for all articles published in that journal.

SB: Another unique model that you have created for managing scientific information was the weekly journal Current Contents *(Garfield, 1979), which included the tables of contents of many scientific journals. Browsing through these tables of contents allowed serendipity. We now perform searches in databases, but we don't browse (through shelves of books, tables of contents of journals). Are we missing a lot of information by just performing searches and not browsing?*

EG: The *Current Contents* has been displaced by the publisher alerts you can now receive gratis from journals. You can receive these alerts by email and browse them to see the articles published in the latest issues. You can also receive citation alerts from Thomson Reuters and others based on your personal profile.

SB: Your essays in Current Contents *have educated scientists and librarians on how to manage scientific information. This task has become much more difficult today. How do you manage information for your personal use?*

EG: I don't do research anymore. I follow the Special Interest Group for Metrics (SIG-Metrics) of ASIS&T (the Association for Information Science and Technology). You don't need to be a member of ASIS&T to access this Listserv. I also follow other Listservs such as CHMINF-L (the Chemical Information Sources Discussion List).

SB: Managing research data is becoming a major challenge for researchers and users. How will the organization, retrieval, and management of literature and research datasets be integrated in the future?

EG: If everything becomes open access, so to speak, then we will have access to, more or less, all the literature. The literature then becomes a single database. Then, you will manipulate the data the way you do in Web of Science, for example. All published articles will be available in full text. When you do a search, you will see not only the references for articles, which have cited a particular article, but you will also be able to access the paragraph of the article to see the context for the citation. By the way, that capability already exists in *CiteSeer*—the computer science database located at Penn State University. When you do a search in *CiteSeer*, you can see the paragraph in which the citation occurs. That makes it possible to differentiate articles much more easily and the full text will be available to anybody.

SB: Some of the new methods for evaluating research use article-level metrics. Is the attention to smaller chunks of information going to challenge the existence of the scientific journal? Are there going to be DOIs for every piece of information published—a blog, a comment, a tweet, a graph, a table, a picture? And how will this tendency for granularity in scientific communication affect the ability of users to manage such diverse information?

EG: Back in the 1950s, Saul Herner published an article, "Technical information: too little or too much?" (Herner, 1956). And the same question is being asked today. You can create a personal profile to receive the information that you need. As long as there are tools for refining searches, I don't think there will be too much information for a scientist who will be focusing on a narrower topic. I don't think, really, that anything has changed in that respect. I still browse the same way, I still search the same way, and I want to go to a deeper level.

SB: There is a lot of hype about altmetrics now, but there are currently mixed opinions on its potential as a credible system for evaluating research. Is academia going to be receptive to these new ways of measuring scientific impact?

EG: This is a big educational problem—to educate administrators about the true meaning of citation impact. Scientists are smart enough to realize the distinctions between IF, h-index, and other indexes for citations. There are now hundreds of people doing bibliometric or scientometric research. I think there will be more refined methods, and I'm not worried about that—as long as people continue to get educated. Proper peer review from people who understand the subject matter will be important. I don't think altmetrics is that much different from the existing metrics that we have. Human judgement is needed for evaluations.

SB: A recent article in Angewandte Chemie *was titled "Are we refereeing ourselves to death? The Peer-Review system at its limit" (Diederich, 2013). It discussed the challenges peer review is presenting to journals, authors, and reviewers. Even highly respected journals such as* Science *and* Nature *had to retract articles that had gone through the peer-review process. Is peer review going to be abolished?*

EG: Joshua Lederberg discussed this problem about 30 years ago—that scientists will publish open access and that we will have open peer review. If you are a scientist and if you are an honest scientist, you don't have to worry about open peer review. People will rely on your reputation. Commentary could come from anywhere and from anyone who wants to contribute to the subject. I don't think peer reviewers will go away. Peer review will evolve to open peer review. I don't believe in the secretive type of peer review. Those scientists who commit fraud will be exposed. If they want to take a chance doing it, that's their problem. But, eventually, they will face worse hurdles for being exposed. And their research won't get cited very much.

SB: *People now work in big collaborations, and it is rare that a single person would publish a paper. How is authorship going to evolve and how will individual authors be recognized?*

EG: It is a very difficult question, but I think the professional societies have to decide about that. For each paper that is published, somebody will have to take responsibility for that paper. At ISI, we processed all authors equally. For the foreseeable future, all authors will have equal weight. The problem remains for administrators, because they want numbers to evaluate people for tenure.

SB: *To follow up on the previous question—researchers who are at a high level, they don't work in the lab. So how could they take responsibility for a paper when someone else has done the actual work?*

EG: That depends on the author. It comes down to personal judgement.

(All images accompanying this chapter were provided by Eugene Garfield and reproduced with his permission)

More information about Eugene Garfield is available at www.garfield.library. upenn.edu.

References

Brin, S., Page, L., 1998. The Anatomy of a Large-Scale Hypertextual Web Search Engine. Retrieved September 8, 2014, from http://infolab.stanford.edu/~backrub/google.html.

Diederich, F., 2013. Are we refereeing ourselves to death? The peer-review system at its limit. *Angew. Chem. Int. Ed.* 52 (52), 13828–9. http://dx.doi.org/10.1002/anie.201308804.

Garfield, E., 1964. "Science Citation Index"—new dimension in indexing. *Science* 144 (3619), 649–54. http://garfield.library.upenn.edu/essays/v7p525y1984.pdf.

Garfield, E., 1979. *Current Contents*. Its impact on scientific communication. *Interdiscip. Sci. Rev.* 4 (4), 318–323. http://garfield.library.upenn.edu/essays/v6p616y1983.pdf.

Garfield, E., 2006. The history and meaning of the Journal Impact Factor. *JAMA* 295 (1), 90–93. http://garfield.library.upenn.edu/papers/jamajif2006.pdf.

Herner, S., 1956. Technical information: too much or too little? *Sci. Mon.* 83 (2), 82–6. http://garfield.library.upenn.edu/papers/herner1956.pdf.

What it looked like to work at the Institute for Scientific Information (ISI): interview with Bonnie Lawlor

13

These are excerpts from the following previously published interview:

Baykoucheva, S. (2010) From the Institute for Scientific Information (ISI) to the National Federation of Abstracting and Information Services (NFAIS): Interview with Bonnie Lawlor. *Chemical Information Bulletin*, *62*(1), 17–23.

A chapter on the history of ISI by Bonnie Lawlor (Lawlor, 2014) was published recently in the book *The Future of the History of Chemical Information* (McEwen and Buntrock, 2014).

Bonnie Lawlor

Bonnie Lawlor is the executive director of NFAIS, a membership association for organizations that aggregate, organize, and facilitate access to authoritative information. Bonnie was the executive vice president of the Database Publishing Division at the Institute for Scientific Information (ISI) (now Thomson Reuters, Healthcare & Science), where she was responsible for product development, production, publisher relations, editorial content, and worldwide sales and marketing of all of ISI's products and services.

Svetla Baykoucheva: You have held a number of executive positions in different companies and nonprofit organizations and you have served as an elected official in the American Chemical Society (ACS). It seems that all the organizations that you have been affiliated with professionally have something in common—they are all related to scientific information and scientific publishing. How did you come to this field, what triggered your interest in it, and what were the main factors that have influenced your career (e.g. education, chance, timing, etc.)?

Bonnie Lawlor: …I fell into the field of scientific publishing quite unintentionally. Immediately after college, I went to the University of Pennsylvania to study for my PhD. Upon completion of my coursework, I left to find a job as I had become engaged to a Vietnam War veteran who wanted to complete his college degree. With only a bachelor's degree in chemistry, the opportunities were less than exciting, plus I was uncertain as to whether or not a laboratory career was really for me. I saw an advertisement for a chemical indexer in the now defunct *Philadelphia Bulletin*. I had no idea what being a "chemical indexer" actually entailed, but I interviewed, was tested, and was offered the position at the Institute for Scientific Information (ISI). After two years, I was hooked. ISI was, at that time, small, entrepreneurial, and very interesting. Plus I was able to use my education and love of the theory of chemistry without having to spill chemicals (which I had been known to do!). Ultimately, I became involved with other areas of the company—*Current Contents*, the citation indexes, etc.—and was caught up in the industry transition from print to electronic publications. An exciting era only made more so by the introduction and evolution of the web!

SB: Being executive vice president of the Database Publishing Division of the Institute for Scientific Information (ISI is now Thomson Reuters, Healthcare & Science) and being responsible for so many areas (product development, production, publisher relations, editorial content, and worldwide sales and marketing of all of ISI's products and services) could be a daunting responsibility. What imprint, do you think, your work has made on ISI's success and image?

BL: Over the 28-year span that I spent at ISI, I would perhaps choose a few "turning points" where I know that I had an impact on the outcome and the ultimate shaping of the company. The first is regarding ISI's chemical information products. *Index Chemicus*, a weekly alert to new chemical compounds, was launched by Dr. Garfield in the early 1960s before I joined the company. It was not a popular move and three vice presidents even left the company, partially due to this initiative that they perceived as being risky. In 1982, the entire chemistry product line was made a separate division under my leadership, with the directive to make it work. We were responsible for product development, production, sales, and marketing. … By the mid to late 1980s, the entire abstracting and indexing community faced another challenge—how to adapt its print products and services to the newly emerging digital environment sparked in 1981 by the launch of personal computers and fueled by the emergence of the CD-ROM and diskette distribution media. We were very fortunate. We had been creating electronic versions of all of our citation indexes, *Current Contents*, and the chemical products as a by-product of computerized production that most major A&I [Abstracting & Indexing] services had adopted in the 1960s. The issue was to take the data already available on magnetic tape and make it compatible with the new platforms. Change is not easy and it took some doing to convince staff (and in some cases management) that digital was the future. … Within 2 years, 20% of our print base had converted to the new format.

… The second decision that I was able to get approved was to add English language author abstracts to ISI products. Up until this time, they were only included in the print issues of *Index Chemicus* and *Current Chemical Reactions*, and I believed that they were an essential addition to our new electronic offerings… In general, I would say that the combination of my fiscal responsibility and love of ISI together was a great foil to Dr. Garfield's creativity and drive. Throw in the unbelievable genius of people such as Irv Sher, George Vladutz, and Henry Small and the work ethic and loyalty of hundreds of employees who were devoted to the company—ISI became a major force in the information community. I was just one of many and I am grateful that I had the opportunity to be part of the unique ISI family.

SB: What did it take to work and succeed in an environment (such as the one at ISI at that time) that was so innovative, dynamic, and competitive—and dominated by a mythological figure such as Eugene Garfield? Could you tell us what your first encounter with Dr. Garfield was?

BL: …when I joined ISI, it was relatively small and very entrepreneurial. We all were made to feel that we were part of the creation of something of value. When a customer wrote to tell Dr. Garfield that a product or service solved a problem, he let us know (of course, we also heard all of the complaints). It was a truly nourishing environment… No matter what your gender, color, or educational status—if you had an idea, Dr. Garfield was willing to hear it. It was an environment that offered great opportunity if you were creative and willing to work hard. It was also a crazy place to work—perhaps due to the culture of the late 1960s and early 1970s. People parked their motorcycles by their desks. The work dress ranged from normal to eccentric. I remember one person wore baby-doll pajamas to the office and one executive always wore a small teddy bear on his belt (the same two people streaked at one of the company parties!). When my boss complained about the length (or lack thereof) of miniskirts, the corporate (unofficial) response was that the only dress code requirement was shoes!…

I suspect the ISI environment was a combination of the times and the personality of our corporate leader. I still smile about my first encounter with Dr. Garfield. Every day, the coffee shop in the lobby of our building sent a cart to each floor in midmorning and afternoon so that everyone could get a snack. While I waited in line by the elevators to get my caffeine fix in the early days of my employment, a rather strange vision emerged from the elevators wearing a gray jacket with a fur collar and wild hair reminiscent of Albert Einstein. I asked the person behind me who it was (I thought perhaps he was a handyman). When the laughter subsided, I was told the vision in question was Dr. Garfield. Ultimately, I came to know, respect, and occasionally fear him. I learned so very much from him—the importance of such things as quality, responsiveness to customers, innovation, and being a professional. Even though we competed with the American Chemical Society, he made sure that we were active in the ACS—particularly in what is now the Division of Chemical Information. He said that we were chemists and should actively promote the profession… In retrospect, I could not have had a better mentor. We still keep in touch and I treasure our relationship.

SB: You have been involved in database publishing for a long time. How do you see the future of the secondary publishers? How will models such as Google Scholar that rely on parsing the full text of documents affect the commercial databases and in what respect? How will services such as PubChem affect the commercial vendors of chemical property information?

BL: I believe that the current climate of change in scholarly communication will impact all publishers, both primary and secondary. ... The Internet, like the printing press before it, has created an information revolution that is generating new forms of scholarly communication and publishing ... A&I services as we know them began to emerge in the early 1800s, when there were approximately 300 scientific journals. Since then, their purpose has never changed: they play an essential role in allowing scholars to navigate masses of information with relative ease. The bibliographic pointers such as keywords, subject indexes, authors, titles, etc., facilitate the discovery of information; abstracts allow the evaluation of a document's relevance to one's research; and links— either a bibliographic reference or, in today's world, an electronic link—allow retrieval of the full text. And as over the years these services build a body of information, they serve as the continuum between past, current, and future scholarly thinking upon which all human knowledge is built. This is the essential role that organizations such as CAS and ISI play even today. They began when scholarly communication was print-based and they have adapted; we now progress through a transition consisting of both print and digital media. You have raised two issues, the first dealing with Google Scholar (and this can be extended to all free information on the web) and the second dealing with scholarly information services that are available from the government or have been established using an open-access business model such as the Public Library of Science (PLoS). ... But when researchers become involved in a specific project, they turn to the more traditional services offered by their libraries or information centers in order to obtain their information, and they do so for two reasons:

1. They know that these services cover the source material in which the vast majority of scientists and scholars publish (Google Scholar does not).
2. They know that these services provide authoritative, reliable content (all Google content is not reliable) ...

Having said that, I do believe that the well-established A&I services are vulnerable if they do not pay attention to the new forms of scholarly communication. Their charter is to facilitate the discovery of and access to scholarly and scientific information. As the primary basis of that communication (journals) evolves into a more dynamic, online, collaborative "conversation," they must adapt their services to capture and preserve the content of the conversation... But they must ensure that they deliver products offering ease of access to all the available information that is needed by their particular user base—no matter what the source. Traditional A&I services have the knowledge and expertise to be the A&I services of the future. But they must embrace the new forms of scholarly communication today, not ignore them and not wait and see ...

References

Lawlor, B., 2014. The institute for scientific information: a brief history. *ACS Symp. Ser. 1164* (Future of the History of Chemical Information), 109–26. http://dx.doi.org/10.1021/bk-2014-1164.ch007.

McEwen, L.R., Buntrock, R.E. (Eds.). (2014). *The Future of the History of Chemical Information* (Vol. 1164): American Chemical Society.

Measuring attention: social media and altmetrics

14

> Very simple was my explanation, and plausible enough—as most wrong theories are!
>
> *H.G. Wells* (The Time Machine, *1895*)

14.1 Introduction

We have heard of alternative music, alternative medicine, alternative energy, and many other "alternative" things. There are even "alternative mutual funds." In the academic and publishing world, we now often hear about alternative metrics, or altmetrics. Why are so many people and organizations interested in this new field? Altmetrics monitors attention to scholarly output in the form of mentions on social media sites, scholarly activity in online libraries, and reference managers and comments posted on scientific blogs. Who talked about that paper? How did they engage? How can the impact of their activity be measured?

Traditional methods of evaluating the impact of research were based on citations in peer-reviewed journals. The journal impact factor (IF), the most widely used measure of scientific impact, was developed in a print environment, but other alternative indicators have also been studied by such fields as webometrics and bibliometrics (Baynes, 2012; Corrall et al., 2013).

After scholarly communication shifted mostly to online, there was a need for a new approach to evaluate research. As stated in a document published by the NISO Alternative Assessment Metrics (Altmetrics) Project, "While citations will remain an important component of research assessment, this metric alone does not effectively measure the expanded scope of forms of scholarly communication and newer methods of online reader behavior, network interactions with content, and social media" (NISO, 2014).

Authors like to know who is looking at their works and what other people think about them. Why would you make the effort to publish if you do not care what others think about your work or if they see the impact of it? This interest in other people's opinions has reached new highs with the development of advanced Internet technologies and the emergence of social media. What people are saying about an article is being viewed as some kind of an indicator of interest. Many journals now display alternative metric information at article level. Criteria such as the number of "likes" and the number of Twitter followers are viewed by some as measures of research impact (Holmberg and Thelwall, 2014; Kraker, 2014; Osterrieder, 2013; Sud and Thelwall, 2014). An article that examined how often Twitter was used to disseminate information about journal articles in the biomedical field concluded that the correlation between tweets and citations was very low (Haustein et al., 2014b).

The open-access movement, widespread use of social media, availability of free content, and new forms of scholarly output such as datasets led to the development of new alternative metrics to analyze research. The tools and the data that can be collected using them are grouped under the collective term "altmetrics."

14.2   Measuring attention

The term "altmetrics" is an abbreviation of the phrase "alternative metrics" (Galligan and Dyas-Correia, 2013). This new field and its supporters have the ambitious goal of providing an alternative to or an enhancement of traditional citation metrics by measuring scholarly interactions taking place mainly in the social media (Galloway et al., 2013; Hoffmann et al., 2014; Kwok, 2013; Piwowar, 2013; Piwowar and Priem, 2013; Sud and Thelwall, 2014; Taylor, 2012; Viney, 2013; Wang et al., 2013; Wilson, 2013). These interactions may take the form of article views, downloads, and tweets. They could also be collaborative annotations using such tools as social bookmarking and reference managers and comments on blog posts. Altmetrics companies obtain data from many different sources and gather metrics for such digital artifacts as articles, blog posts, book chapters, books, cases, clinical trials, conference papers, datasets, figures, grants, interviews, letters, media, patents, posters, presentations, source code, theses/dissertations, videos, and even web pages.

Plum Analytics (Plum Analytics, 2014a,b), a major player in the field of altmetrics, separates collected data into the following five categories:

- Usage (e.g. downloads, views, book holdings, ILL, and document delivery).
- Captures (favorites, bookmarks, saves, readers, and groups).
- Mentions (citations from blog posts, news stories, Wikipedia articles, comments, and reviews).
- Social media (tweets, +1's, likes, shares, and ratings).
- Citations (retrieved from publicly available sources such as PubMed, Scopus, and patent databases).

14.3   Altmetrics companies, applications, and tools

Companies and organizations involved in altmetrics collect data from many sources, including social media outlets (Baynes, 2012; Brown, 2014; Buschman and Michalek, 2013; Cheung, 2013; Haustein et al., 2014a; Konkiel, 2013; Kwok, 2013; NISO, 2014; Piwowar, 2013; Piwowar and Priem, 2013; Sud and Thelwall, 2014; Thelwall et al., 2013; Wang et al., 2013; Wilson, 2013). While most altmetrics companies are using similar metrics, categories, and sources of data to those shown above for Plum Analytics, there are also some differences in the approaches used by these companies. This section provides information about the applications and tools currently used in this field and the major players in it.

14.3.1   Academic Analytics

www.academicanalytics.com

Academic Analytics provides business intelligence data and solutions for research universities in the United States and the United Kingdom that allow them to compare academic institutions and help them better understand their strengths and in which areas they need to improve.

14.3.2   altmetric.com

altmetric.com is a fee-based service that provides altmetrics tools to publishers, institutions, and researchers (Figure 14.1). Publishers subscribing to this service display Article-Level Metrics (ALMs), which draws more visitors to their sites (Huggett and Taylor, 2014).

The company gathers data about mentions of academic papers on social media sites (e.g. Twitter, Facebook, Pinterest, and Google+), science blogs, mainstream media outlets such as *The New York Times* and *The Guardian*, non-English language publications like *Die Zeit* and *Le Monde*, the peer-reviewed site Publons, and information about references saved in collaborative reference managers.

altmetric.com uses rankings for their data analysis. For example, news items have more weight than blogs, and blogs are more highly regarded than tweets. The algorithm also takes into account how authoritative the authors are. Results are presented visually with a donut that shows the proportional distribution of mentions by source type, with each source type displaying a different color—blue (for Twitter), yellow (for blogs), and red (for mainstream media sources). Elsevier displays the altmetric.com donuts for a journal's three top-rated articles on the Elsevier.com homepages of many Elsevier titles. The donut also provides links to news and social media mentions.

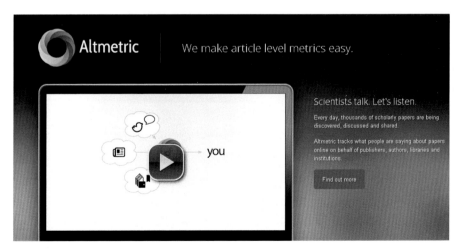

Figure 14.1 The home page of altmetric.com.
Reproduced with permission from altmetric.com.

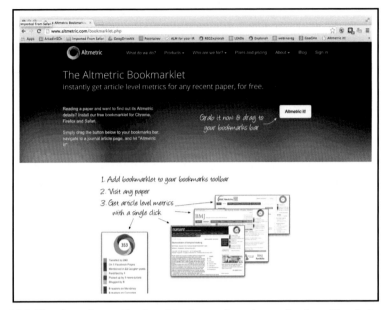

Figure 14.2 The altmetric.com Bookmarklet (www.altmetric.com/bookmarklet.php). Reproduced with permission from altmetric.com.

A free Bookmarklet that can be installed in Firefox, Chrome, and Safari allows users to see details of the use of a particular article (Figure 14.2).

14.3.3 Article-level Metrics (ALMs)

http://article-level-metrics.plos.org

ALMs (Figure 14.3) is a new approach introduced by PLOS to measure the impact of published research at the article level. It looks at an individual article's impact and

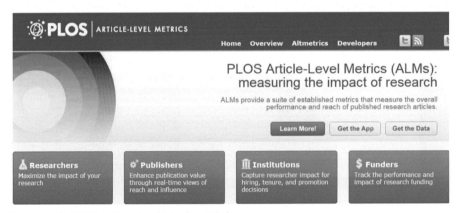

Figure 14.3 PLOS Article-level Metrics website.

separates it from the impact of the journal it was published in. ALMs incorporates altmetrics sources along with traditional measures to present a bigger picture of how individual articles are being discussed and used.

14.3.4   CiteULike

www.citeulike.org

CiteULike is a free collaborative bibliographic management program that allows users to store, organize, share, and discover scholarly references.

14.3.5   Impactstory

https://impactstory.org

Impactstory (Figure 14.4) is an open-source, web-based tool that is focused on researchers. It tracks journal articles, preprints, datasets, presentation slides, research codes, and other research outputs. It is known that Impactstory aggregates data from Mendeley, GitHub, and Twitter, but it does not disclose all its sources. Impactstory was founded by Heather Piwowar and Jason Priem, pioneers in altmetrics and also prolific writers in the area of new publishing models and alternative approaches for evaluating research (Piwowar, 2013; Piwowar and Vision, 2013; Piwowar et al., 2011).

Figure 14.4 The home page of Impactstory (https://impactstory.org). Reproduced with permission from Impactstory.

14.3.6   InCites

http://researchanalytics.thomsonreuters.com/incites/

InCites from Thomson Reuters is using bibliographic and citation data from the Web of Science to provide tools and solutions for assessing scholarly output. InCites measures and benchmarks the research performance of individuals, organizations, programs, and peers.

14.3.7 Mendeley

www.mendeley.com

Mendeley is a free collaborative bibliographic management tool owned by Elsevier. When searching Scopus, users can see the demographics, disciplines, and geographic locations of people who have saved a particular article in Mendeley (Habib, 2014). This kind of crowd sourcing might be of interest to many users, but there are some researchers (especially those working in competitive fields) who are reluctant to use a bibliographic management program that monitors their reading habits.

14.3.8 PaperCritic

www.papercritic.com

PaperCritic allows researchers to get feedback about their articles. Tags, summaries, and in-text notes from an individual researcher's Mendeley library are available to PaperCritic users if they want to rate and review any publication.

14.3.9 Plum Analytics

www.plumanalytics.com/https://plu.mx

Plum Analytics (Plum Analytics, 2014a) (Figure 14.5) is a major player in the field of altmetrics. Recently acquired by the aggregator EBSCO, it is a commercial service targeting mainly libraries.

Figure 14.5 Plum Analytics home page.
Reproduced with the permission of Plum Analytics.

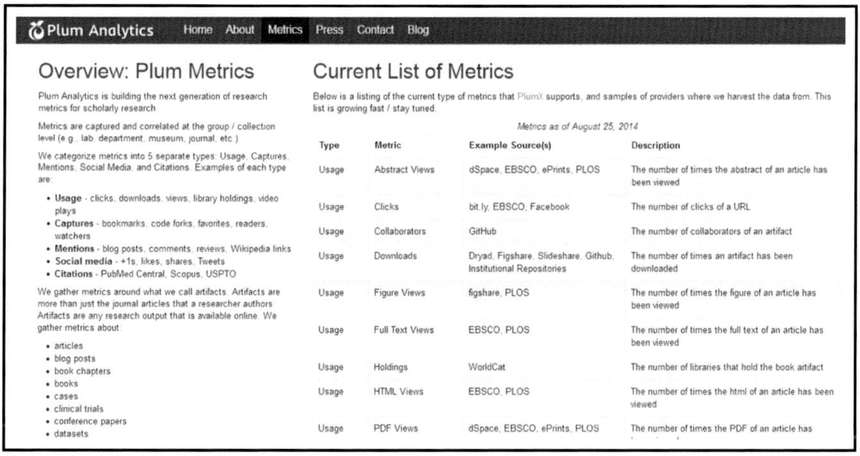

Plum Analytics   Home About **Metrics** Press Contact Blog

Overview: Plum Metrics

Plum Analytics is building the next generation of research metrics for scholarly research.

Metrics are captured and correlated at the group / collection level (e.g. lab, department, museum, journal, etc.)

We categorize metrics into 5 separate types: Usage, Captures, Mentions, Social Media, and Citations. Examples of each type are:

- **Usage** - clicks, downloads, views, library holdings, video plays
- **Captures** - bookmarks, code forks, favorites, readers, watchers
- **Mentions** - blog posts, comments, reviews, Wikipedia links
- **Social media** - +1s, likes, shares, Tweets
- **Citations** - PubMed Central, Scopus, USPTO

We gather metrics around what we call artifacts. Artifacts are more than just the journal articles that a researcher authors. Artifacts are any research output that is available online. We gather metrics about:

- articles
- blog posts
- book chapters
- books
- cases
- clinical trials
- conference papers
- datasets

Current List of Metrics

Below is a listing of the current type of metrics that PlumX supports, and samples of providers where we harvest the data from. This list is growing fast / stay tuned.

Metrics as of August 25, 2014

Type	Metric	Example Source(s)	Description
Usage	Abstract Views	dSpace, EBSCO, ePrints, PLOS	The number of times the abstract of an article has been viewed
Usage	Clicks	bit.ly, EBSCO, Facebook	The number of clicks of a URL
Usage	Collaborators	GitHub	The number of collaborators of an artifact
Usage	Downloads	Dryad, Figshare, Slideshare, Github, Institutional Repositories	The number of times an artifact has been downloaded
Usage	Figure Views	figshare, PLOS	The number of times the figure of an article has been viewed
Usage	Full Text Views	EBSCO, PLOS	The number of times the full text of an article has been viewed
Usage	Holdings	WorldCat	The number of libraries that hold the book artifact
Usage	HTML Views	EBSCO, PLOS	The number of times the html of an article has been viewed
Usage	PDF Views	dSpace, EBSCO, ePrints, PLOS	The number of times the PDF of an article has

Figure 14.6 Screen capture of Plum Analytics list of sources. The full table is available at www.plumanalytics.com/metrics.html.
Reproduced with the permission of Plum Analytics.

While most of the altmetrics companies do not disclose their sources of data, Plum Analytics makes an exception by publishing the full list of metrics it supports (Figure 14.6).

In June 2013, the Alfred P. Sloan Foundation awarded the National Information Standards Organization (NISO) a grant "to undertake a two-phase initiative to explore, identify, and advance standards and/or best practices related to a new suite of potential metrics in the community" (NISO, 2015). The project has the goal of developing and promoting new assessment metrics, such as usage-based metrics and social media statistics.

14.4 Altmetrics and data provenance

If you have posted a work on some publicly available website, you may get regular updates of how many times your work has been viewed and downloaded. These reports come from different places: the original site where your work has been published, altmetrics companies monitoring such outlets, or academic social websites, such as academia.edu and ResearchGate, where you have posted your works.

I had not been paying much attention to such reports, until I suddenly noticed that a report sent to me by an altmetrics company for a tutorial that I had posted on a social website showed around 1000 counts more than a report for the same work that the original company had sent me the day before. The altmetrics company confirmed that their stats had been correct and matched the counts shown on the public site of the company where the original work was posted.

Why would a company send authors results that were significantly different (much lower) from what the same company is displaying publicly? A FAQs page of the company where my tutorial was posted gave some possible explanation for this dis-crepancy. It turned out that some new analytics had been used to correct total counts by excluding bot (search engine crawlers) views that had been counted in the total views. These analytics, I was told by the company, had not been applied to the public site, yet. This explains the difference between the results sent to authors (who received the results normalized for bots) and those displayed for the general public and that were collected by the altmetrics companies. There is also no way to know who is view-ing or downloading the works—whether these are people who are really interested in the works, or whether the authors themselves are repeatedly viewing and downloading their own works to increase their counts.

14.5 Conclusion

While it takes years for traditional citations to accrue, the number of downloads, views, and mentions in social media is reported in a matter of days, even hours (Wang et al., 2013). Researchers engaging in social networks now rely significantly on rec-ommendations from their peers about newly published articles. They can see what their peers are finding, saving, and bookmarking. There is a general consensus that having an article mentioned in social media could lead to future citations.

With budgets tightening and funding sources becoming more limited, scientific research is becoming very competitive. When applying for grants, researchers need to show that their research will have an impact. They can demonstrate such potential impact for their past work, but not for recently published papers that might be the most relevant to the grant proposal. If researchers could show that their recent work is gen-erating a lot of interest, this could give them an advantage in getting funded.

Academic libraries are now looking at this new field as an opportunity to play a more prominent role in their organizations (Galloway et al., 2013). They organize seminars, invite representatives from altmetrics companies, and engage in different initiatives aimed at evaluating opportunities for adopting alternative metrics and at the same time educating researchers about this new field (Brown, 2014; Corrall et al., 2013; Lapinski et al., 2013; Wilson, 2013). Professional organizations such as the American Chemical Society devote technical sessions and panel discussions to al-ternative metrics, thus allowing subject librarians and researchers to get acquainted with the field. The altmetrics companies targeting academic institutions require paid subscriptions, and librarians are in a good position to get involved in the selection, introduction, and promotion of such services.

There are significant differences between the disciplines in using social media, which affects whether alternative metrics could be applied for certain fields (Haustein et al., 2014a; Holmberg and Thelwall, 2014; Liu et al., 2013; Zahedi et al., 2014). Scholars in the humanities and the social sciences are very interested in altmetrics. The research output in these disciplines is mainly in the form of books and book chap-ters, and traditional citation analysis, which has been applied most often to disciplines where journal articles have been the main research product, does not serve them well when they are considered for promotion.

It is difficult to predict how the field of altmetrics will develop in the near future, but one thing is certain—its acceptance will be much slower in disciplines where citations in peer-reviewed journals with high IF are major criteria for researchers' promotion. Finding a correlation between citations and counts of downloads, bookmarks, and tweets has been attempted in recent years, but there is no definitive conclusion whether such correlation exists (Baynes, 2012; Zahedi et al., 2014).

The provenance of data and interpreting the collected data will be the most important and challenging issues confronting altmetrics companies in the future. Counts mean nothing, unless they can be interpreted. Altmetrics companies need to explain what the value of their services is. Although altmetrics offers an interesting insight into how scholarly output attracts attention, it will complement rather than replace traditional methods such as citations in peer-reviewed journals (Brody, 2013; Brown, 2014; Cheung, 2013).

References

Baynes, G., 2012. Scientometrics, bibliometrics, altmetrics: some introductory advice for the lost and bemused. *Insights* 25 (3), 311–315. http://dx.doi.org/10.1629/2048-7754.25.3.311.

Brody, S., 2013. Impact factor: imperfect but not yet replaceable. *Scientometrics* 96 (1), 255–7. http://dx.doi.org/10.1007/s11192-012-0863-x.

Brown, M., 2014. Is altmetrics an acceptable replacement for citation counts and the impact factor? *Ser. Libr.* 67 (1), 27–30. http://dx.doi.org/10.1080/0361526X.2014.915609.

Buschman, M., Michalek, A., 2013. Are alternative metrics still alternative? Retrieved September 3, 2014, from http://www.asis.org/Bulletin/Apr-13/AprMay13_Buschman_Michalek.html.

Cheung, M.K., 2013. Altmetrics: too soon for use in assessment. *Nature* 494 (7436), 176. http://dx.doi.org/10.1038/494176d.

Corrall, S., Kennan, M.A., Afzal, W., 2013. Bibliometrics and research data management services: emerging trends in library support for research. *Libr. Trends* 61 (3), 636–74.

Galligan, F., Dyas-Correia, S., 2013. Altmetrics: rethinking the way we measure. *Ser. Rev.* 39 (1), 56–61. http://dx.doi.org/10.1080/00987913.2013.10765486.

Galloway, L.M., Pease, J.L., Rauh, A.E., 2013. Introduction to altmetrics for science, technology, engineering, and mathematics (STEM) librarians. *Sci. Technol. Libr.* 32 (4), 335–45.

Habib, M., 2014. Mendeley Readership Statistics available in Scopus. Retrieved from http://blog.scopus.com/posts/mendeley-readership-statistics-available-in-scopus.

Haustein, S., Peters, I., Bar-Ilan, J., Priem, J., Shema, H., Terliesner, J., 2014a. Coverage and adoption of altmetrics sources in the bibliometric community. *Scientometrics*, 101 (2), 1145–63. http://dx.doi.org/10.1007/s11192-013-1221-3.

Haustein, S., Peters, I., Sugimoto, C.R., Thelwall, M., Lariviere, V., 2014b. Tweeting biomedicine: an analysis of tweets and citations in the biomedical literature. *J. Assoc. Inf. Sci. Technol.* 65 (4), 656–69. http://dx.doi.org/10.1002/asi.23101.

Hoffmann, C.P., Lutz, C., Meckel, M., 2014. Impact factor 2.0: applying social network analysis to scientific impact assessment. Paper presented at the 2014 47th Hawaii International Conference on System Sciences (HICSS), Waikoloa, HI.

Holmberg, K., Thelwall, M., 2014. Disciplinary differences in Twitter scholarly communication. *Scientometrics*, 101 (2), 1027–42. http://dx.doi.org/10.1007/s11192-014-1229-3.

Huggett, S., Taylor, M., 2014. Elsevier expands metrics perspectives with launch of new altmetrics pilots, *Editors' Update*. March 3, 2014. Retrieved June 26, 2014, from http://editorsupdate.elsevier.com/issue-42-march-2014/elsevier-altmetric-pilots-offer-new-insights-article-impact/.

Konkiel, S., 2013. Altmetrics: a 21st-century solution to determining research quality. *Online Searcher* 37 (4), 10–15.

Kraker, P., 2014. All metrics are wrong, but some are useful. Retrieved June 26, 2014, from http://science.okfn.org/tag/altmetrics.

Kwok, R., 2013. Research impact: altmetrics make their mark. *Nature* 500 (7463), 491–3. http://dx.doi.org/10.1038/nj7463-491a.

Lapinski, S., Piwowar, H., Priem, J., 2013. Riding the crest of the altmetrics wave: how librarians can help prepare faculty for the next generation of research impact metrics. *Coll. Res. Libr. News* 74 (6), 292–294.

Liu, C.L., Xu, Y.Q., Wu, H., Chen, S.S., Guo, J.J., 2013. Correlation and interaction visualization of altmetric indicators extracted from scholarly social network activities: dimensions and structure. *J. Med. Internet Res.* 15 (11), e259.

NISO, 2014. Alternative Metrics Initiative. Retrieved September 1, 2014, from http://www.niso.org/topics/tl/altmetrics_initiative/#resources.

NISO, 2015. NISO Alternative Metrics (Altmetrics) Initiative. Retrieved January 8, 2015, from http://www.niso.org/topics/tl/altmetrics_initiative.

Osterrieder, A., 2013. The value and use of social media as communication tool in the plant sciences. *Plant Methods* 9, 26. http://dx.doi.org/10.1186/1746-4811-9-26.

Piwowar, H., 2013. Altmetrics: value all research products. *Nature* 493 (7431), 159. http://dx.doi.org/10.1038/493159a.

Piwowar, H., Priem, J., 2013. The power of altmetrics on a CV. Retrieved September, 2014, from https://asis.org/Bulletin/Apr-13/AprMay13_Piwowar_Priem.html.

Piwowar, H.A., Vision, T.J., 2013. Data reuse and the open data citation advantage. *PeerJ* 1, e175. http://dx.doi.org/10.7717/peerj.175.

Piwowar, H.A., Vision, T.J., Whitlock, M.C., 2011. Data archiving is a good investment. *Nature* 473 (7347), 285. http://dx.doi.org/10.1038/473285a.

Plum Analytics, 2014a. Plum Analytics: Measuring impact. Retrieved from http://www.plumanalytics.com/.

Plum Analytics, 2014b. Plum Analytics: Metrics. Retrieved from http://www.plumanalytics.com/metrics.html.

Sud, P., Thelwall, M., 2014. Evaluating altmetrics. *Scientometrics* 98 (2), 1131–43. http://dx.doi.org/10.1007/s11192-013-1117-2.

Taylor, M., 2012. The new scholarly universe: are we there yet? *Insights* 25 (1), 12–17. http://dx.doi.org/10.1629/2048-7754.25.1.12.

Thelwall, M., Haustein, S., Larivière, V., Sugimoto, C.R., 2013. Do altmetrics work? Twitter and ten other social web services. *PLoS One* 8 (5), 7. http://dx.doi.org/10.1371/journal.pone.0064841.

Viney, I., 2013. Altmetrics: research council responds. *Nature* 494 (7436), 176. http://dx.doi.org/10.1038/494176c.

Wang, X., Wang, Z., Xu, S., 2013. Tracing scientist's research trends realtimely. *Scientometrics* 95 (2), 717–29. http://dx.doi.org/10.1007/s11192-012-0884-5.

Wilson, V., 2013. Research methods: altmetrics. *Evid. Based Libr. Inf. Pract.* 8 (1), 126–128.

Zahedi, Z., Costas, R., Wouters, P., 2014. How well developed are altmetrics? A cross-disciplinary analysis of the presence of 'alternative metrics' in scientific publications. *Scientometrics*, 101 (2), 1491–513. http://dx.doi.org/10.1007/s11192-014-1264-0.

Unique identifiers

15

15.1   Introduction

Authors' names may be presented in different ways: full first name and second initial, abbreviated first name and second initial, and abbreviated first name and no middle initial. Chinese names are most often given in full, sometimes with the given name first, sometimes with the family name first, and sometimes initialized. American last names often contain tags such as "Junior" or "Senior." We have to admit that sometimes, authors also bear responsibility for the way they present their names. The transliteration of names from non-Latin alphabets and especially from Asian languages could be difficult and inconsistent. This is a very serious problem, as the number of publications and patents from Asian countries is growing rapidly.

Publishers are using different strategies to handle variations in authors' names. Some databases provide the option of searching for alternative spellings, as discussed below, but their practices are not always transparent—in this process, it is not clear whether the users benefit from such options or if they are being misled. Only the authors themselves would be able to identify which of the different available spellings of their names are associated with their publications.

15.2   Unique author name identifiers

To solve the author name ambiguity problem in scholarly communication, new unique persistent digital identifiers have been introduced. They make it possible to distinguish authors from each other to ensure that researchers' individual work is recognized.

15.2.1   ORCID (Open Researcher and Contributor ID)

www.orcid.org

ORCID (Open Researcher and Contributor ID) provides a persistent digital identifier, allowing the clear identification of any individual researcher. ORCID is a nonprofit organization solving the author name ambiguity problem in scholarly communication by creating a central registry of unique identifiers. Anyone who registers for ORCID is assigned a unique number, which can be integrated in research and funding workflows, manuscript submissions, grant applications, and patent applications. Organizations may link their records to ORCID identifiers and register their employees and students to obtain ORCID numbers.

ORCID ID example: 0000-0001-6298-2335

15.2.2 International Standard Name Identifier (ISNI)

www.isni.org

ISNI is a global standard persistent unique number assigned by an international body. It is used to uniquely identify contributors to creative works, including researchers, inventors, writers, artists, performers, producers, publishers, and those involved in works distribution.

15.2.3 ResearcherID (Thomson Reuters)

ResearcherID.com is a free resource for the international scholarly research community. After registering, researchers are assigned an individual ID number that stays with them over the course of their careers, regardless of name changes or change in institution affiliation. Authors can create and update their profile, build their "My Publications" list by uploading works that they authored from Web of Science, manage their publication list with EndNote, make their profiles public or private, view citations in Web of Science, find collaborators, and review publication lists. Institutions can designate an administrator to update ResearcherID publication lists of their researchers.

ResearcherID provides tools for a visual analysis of research networks based on subject category, country, institution, author name, publication year, and geographic location. These tools allow authors to visually explore who is citing their published works (Figure 15.1), the subject areas of the citing papers (Figure 15.2), and from which geographic locations they come from (Figure 15.3).

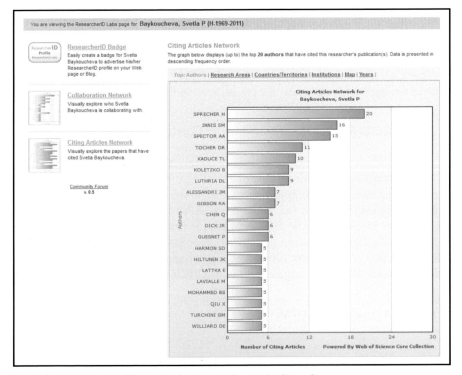

Figure 15.1 ResearcherID shows who has cited an author's works.

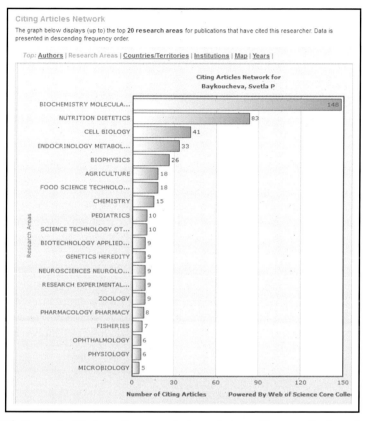

Figure 15.2 ResearcherID shows the areas of science that are citing an author's works.

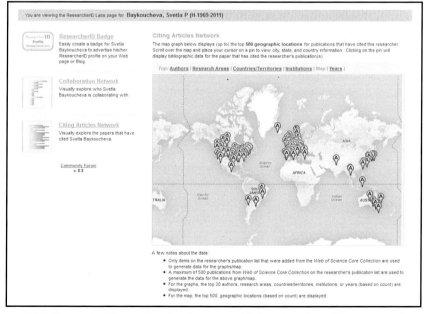

Figure 15.3 ResearcherID map showing the geographic locations for publications that have cited an author's works.

ResearcherID example: B-9809-2012

All screen captures from ResearcherID are reproduced with the permission of Thomson Reuters.

15.3   Handling of author names by publishers

15.3.1   SciFinder

SciFinder searches for alternate spellings of authors' names. As shown in Figure 15.4, an author search produced eight different variations of my name. Some of the variations were caused by the fact that SciFinder simultaneously searches two databases—the Chemical Abstracts (CAplus) and MEDLINE databases—and some of the journals that they cover are the same. The variations included different spellings of my last name; first and middle initials; full or abbreviated first name; and the presence or absence of middle initial. I found out that some of my publications were indexed exactly as they were published in the article, while other references of the same article had my first name abbreviated to an initial. While it was not difficult for me to see that all these were variations of my name, it might not be possible for anyone else to make such distinction.

Figure 15.4 Variations of the author's name, as shown in SciFinder.

Screen capture used with the permission of the Chemical Abstracts Service (CAS), a division of the American Chemical Society.

15.3.2 Scopus Author Identifier

Scopus Author Identifier allows grouping documents published by the same author under a single identifier number, using an algorithm that matches author names based on their affiliation, address, subject area, source title, dates of publication citations, and coauthors. It also takes into consideration last name variations, all possible combinations of first and last names, and the author name with and without initials. The result is that searches for a specific author include a preferred name and variants of the preferred name.

Figure 15.5 shows the Scopus interface when searching by author's name.

The author details page (Figure 15.6) provides information about authors who have more than one document matched to them in Scopus. The information includes the author's publishing and citation information, name variants, and the author's unique identifier number.

On the author details page, users can see the individual author's documents. From this page, authors can request corrections to their names and other details. They can also evaluate an author's performance (e.g. from the h-index), view an author's affiliations, find and group potential author matches, and add author details to ORCID.

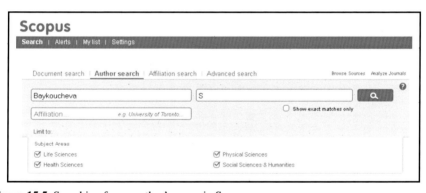

Figure 15.5 Searching for an author's name in Scopus.

Figure 15.6 The author detail page allows the selection of any individual author's different "identities." Authors can use this page to request the merging of their multiple profiles.

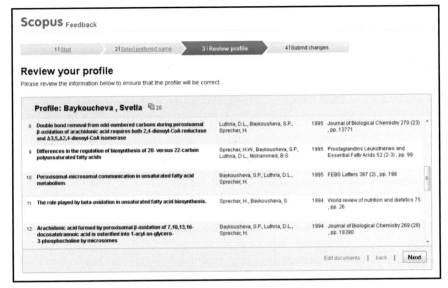

Figure 15.7 Reviewing an author's profile in Scopus.

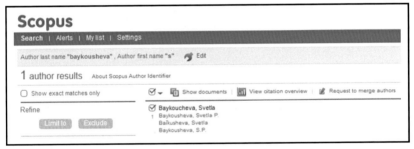

Figure 15.8 Variations of an author's name can be matched to a single name chosen by the author for the Scopus profile.

Authors can report errors or provide other feedback using the author feedback form. An author can review her profile (Figure 15.7) and also select which variation of her name should be the main form for the profile (Figure 15.8). Variations can be added; for example, when I found out that in one of my publications my last name had been misspelled, I contacted Scopus and requested that this variation of my name be added to my profile.

All screen captures from Scopus are used with the permission of Elsevier.

15.3.3 Web of Science

Web of Science also allows authors to manage their names. They have to contact the publisher (Thomson Reuters) to do so. Searches for authors' publications in WoS can also be carried out using ORCID or ResearcherID, discussed earlier in this chapter.

15.4   Other unique identifiers

15.4.1   Digital object identifier (DOI)

www.doi.org

Most journal articles can now be identified by a digital object identifier (DOI), issued by CrossRef (www.crossref.org), a nonprofit organization, in which most scholarly publishers participate as members. (CrossRef is discussed in more detail in Chapter 2 of this book.) A DOI is used to identify a digital object such as a journal article, a book chapter, or even a dataset. Each DOI is associated with the metadata that identify the content of the item and points to its location on the Internet. DOIs are similar to International Standard Book Numbers (ISBN), allowing persistent and unique identification of a publication (or even part of a publication), independently of its location. DOIs are also used for other kinds of content. DataCite (www.datacite.org) is another nonprofit organization that can issue DOIs, and it does so mainly for datasets. Other unique identifiers include PubMed ID, ArXiV ID, and PubMed Central ID. Material posted in university repositories may also be assigned a DOI.

15.4.2   Identifiers for chemical names

The International Chemical Identifier (InChI) is a textual identifier for chemical substances to encode molecular information and to facilitate the search for such information in databases and on the web (Heller et al., 2013). It was initially developed by IUPAC (the International Union of Pure and Applied Chemistry) (www.iupac.org) and NIST (the National Institute of Standards and Technology) (http://nist.gov), using a format and algorithms that are nonproprietary. SMILES (Simplified Molecular Input Line Entry System) is another chemical nomenclature and data-exchange format that is widely used. Another unique identifier for chemical compounds is the CAS Registry Number, discussed in more detail in Chapter 5 of this book.

15.5   Conclusion

After registering for an author identifier, researchers are assigned an individual ID number that stays with them over the course of their careers, regardless of how their names or institution affiliations change. While anybody can get an ORCID number, a ResearcherID number can be assigned only to those who are already authors of scientific publications, and their publications are covered in Web of Science. Now many publishers provide the option to search for authors' publications using their ORCID and ResearcherID numbers instead of their names.

References

Heller, S., McNaught, A., Stein, S., Tchekhovskoi, D., Pletnev, I., 2013. InChI - the worldwide chemical structure identifier standard. *Journal of Cheminformatics 5*, 7. http://dx.doi.org/10.1186/1758-2946-5-7.

Epilogue: creating an information-literate generation of scientists

"Are you talking about EndNote?" asked the driver of the shuttle bus. "That was the most useful thing I have learned in this school." Most of the shuttle bus drivers at the University of Maryland are students. The driver who asked this question had overheard a conversation I was having with someone on the bus. For several years, I taught EndNote workshops for the whole campus, and only hours after the announcements went out, new applicants had to be placed on waiting lists. Most of those who signed up for these classes were faculty and graduate students. In these workshops, attendees not only were able to learn the basics of a bibliographic management program but also searched databases in their own field, exported references to their EndNote libraries, and inserted citations in papers they were writing. I have applied the same model of integrating bibliographic management into information literacy classes taught in science courses. The results and the positive feedback from both students and teaching faculty have exceeded my expectations.

Years after I stopped offering these workshops, requests to offer them again continued to come. A recent e-mail from a student shows how much students appreciated learning bibliographic management and the way I introduced them to it:

> I was in professional writing English class a year ago and for one of our meeting periods, we came to your workshop at McKeldin library. I am now writing a paper for my Conservation Biology class and just wanted to thank you for putting the time into making that handout. I had never heard of EndNote Web before attending your workshop, but now wish I had much sooner because it is a wonderful resource. It has been a while since the workshop, and honestly a while since I have had to write any large research papers. Due to this, I had forgotten a lot of what I had learned in the workshop. Thankfully, I came across a few old papers that I saved, including the Mastering EndNote Web Handout. It has been a huge help, so I just wanted to reach out and thank you for making it.

After the era of the EndNote workshops was over, I became interested in electronic laboratory notebooks (ELNs). This technology is now widely used in industry, but academia has been slow to adopt it. In my "previous life," as a lab-bench researcher, I struggled with paper notebooks, just as researchers are struggling with them even today. Several workshops on ELNs for campus were followed by a pilot project to introduce the use of ELNs in a chemistry course. In a few months, I will be participating in an effort funded by a grant to introduce ELNs in two chemistry courses—a small one (with 50 students) and a large one with 400 students.

While these workshops attracted many people, some of my colleagues were wondering why I was doing that at all. Library administrators were asking me why I was teaching so much. I am glad I taught them, because they led to my involvement in

Managing Scientific Information and Research Data

many other information-literacy projects, writings, presentations, invitations for collaborations, and interactions with researchers and students. The workshops made me the "go-to" person for questions about bibliographic management tools and ELNs. Last year's successful experiment in teaching information literacy sessions in a large chemistry course with almost 450 people will be replicated this year with a freshman chemistry course with 800 students.

It is amazing how quickly students, even in early chemistry courses, are able to learn to search complex chemistry databases. They find it fascinating to draw a molecular structure in SciFinder or Reaxys and discover the chemical compound this structure corresponds to. A student told me that when he was drawing a structure in these databases, he felt as if he was doing real research in the lab. Another student (in a 200-level chemistry course) was similarly impressed:

> I thought this class and assignment were absolutely spectacular and critical in the develop[ment] of our science careers!!! Wow, how I wish I knew about these resources earlier!!! Show this to high school students, I am curious to the amount of hidden talent there when given these resources!!!! Thank you! This needs to be done in more science classes.

I am now going to say something that some of my colleagues (and library administrators) will not agree with: I see information literacy, in the context of how it is discussed in this book, as the most important and exciting area for librarians to be involved in. Look at the opportunity to teach 800 students how to search sophisticated chemistry databases, find literature and property information, and use bibliographic management tools to write papers, dissertations, and books. And then compare it with another possibility being promoted now in academic libraries—helping individual researchers prepare data management plans (there are templates for this now) or manage their research data (if they trust subject librarians who have no subject background to do it). What would be more valuable to the university (and to society, in general)? I realize that this could sound like heresy coming from someone who has just written a chapter on eScience and academic libraries. Time will show how priorities for librarians will shift in the near future and whether my enthusiasm for information literacy will subside.

Teaching information literacy is like writing or making a presentation—you could do it in a boring way or make it exciting. Teaching is helping students understand and remember and also become curious. They will forget much of what they have learned in their science courses, but if you have raised their curiosity and have showed them how to find and manage scientific information, they will be able to find their way. And how you do it makes a difference. Just focus on the content, not letting the euphoria about technology take you away from what is important for students to know.

Managing scientific information and research data constitutes an important part of a researcher's professional life. Whether someone will be considered information-literate depends to a great extent on what type of information is valued in that person's particular environment and area of study. A person considered to be well-informed in one culture may not be perceived this way in another. With access to so much information and technology, it is only a matter of personal effort for an individual to become information-literate and learn how to manage this information.

Index

Note: Page numbers followed by *f* indicate figures.

A

Academia.edu, 16, 133. *See also* Altmetrics; Social media
ACS Division of Chemical Information, 4, 97, 98, 125. *See also Chemical Information Bulletin*
Alternative metrics, 9, 73, 127, 128, 134. *See also* Altmetrics
Altmetric.com, 129–130. *See also* Altmetrics; Social media
Altmetrics, 120, 127–135. *See also* Alternative metrics; Social media
American Chemical Society (ACS), 3, 10, 36, 87, 97, 98, 123–124
Article-level metrics, 14, 120, 129, 130–131. *See also* Alternative metrics; Altmetrics
Articles, rejection of, 21, 23, 30, 31. *See also* Editors; Fourkas, John, interview with
ArXiv, 12, 40, 143
Assignments, sample questions for, 58–62
Association of Research Libraries (ARL), 3, 12, 15, 79. *See also* eScience
Author affiliations, 34, 40, 138, 141
Authorship, 4, 19–20, 22–23, 24, 121. *See also* Bader, Alfred; Ethical issues in science; Gordin, Michael
Authors' names, 36, 137–140, 141, 141*f*. *See also* Unique identifiers

B

Bader, Alfred, 4, 19
Beilstein Handbook of Organic Chemistry, 4–5, 35–36. *See also* Reaxys
bepress, 12
Bibliographic management, 5–6, 37, 44–50, 51*f*, 52*f*, 54*f*, 66, 73, 103, 132, 145–146
Bibliometrics, 115, 127
BioMed Central, 12

Biomedical information, 11, 14, 34–35, 37
BioRxiv, 12
Blogging, 14, 15–16, 127, 129–130. *See also* Social media
Boukacem-Zeghmouri, Chérifa, interview with, 65–69

C

CAplus, 36, 37–38, 38*f*, 40, 140
Careers, 1, 2–3, 15, 37, 97, 116, 123–124
CAS Registry Number, 36, 143
Chemical Abstracts Service (CAS), 36, 40, 56*f*, 59, 60, 126, 140, 143
Chemical information, 4–5, 6, 97–101, 124
Chemical Information Bulletin, 4, 19, 98, 123
Chemical properties, 33, 35, 48, 125
Cheminformatics, 15, 97, 98
Chemistry Central, 4, 11
Chemists, 4–5, 11, 16, 19, 91, 92, 93, 94, 97–100
CHMINF-L, 97–98, 119. *See also* Wiggins, Gary, interview with
CiteULike, 46, 131
Citing previous work, 23–24, 26, 47, 107–109, 138, 139*f*
Clearing House for the Open Research of the United States (CHORUS), 12
CODATA, 74, 77
Committee on Publication Ethics (COPE), 26
Comparing databases, 37, 47
Compendex, 34
Council of Science Editors (CSE), 26
CrossCheck, 26
CrossRef, 12, 13, 26, 112, 143
Culture, organizational, 2, 9, 11, 15, 44, 75, 82, 94, 98–99, 125, 146
Current Contents, 1, 6, 103, 116–117, 116*f*, 117*f*, 118, 119, 124

D

Data. *See also* Electronic laboratory
notebooks (ELNs); eScience
citing of, 74
curation, 72–73, 77, 79–80
management, 5, 6, 13, 72, 73, 77, 78–80,
81–82, 91, 92, 146
management plans, 78–79, 81, 146
preservation, 73, 76–78, 80
provenance, 72, 86, 133–134
repositories, 76–78, 93–94
sharing, 10, 75–76
standards, 73–74
Databases, 33, 34–36, 37–38, 40, 66, 67–68,
103, 108, 115, 119, 120
Databib, 77, 78
DataCite, 77, 78, 143
DataNet, 77
DataONE, 77
Datasets, 13, 14, 15, 71–73,
74, 76, 78, 79–80, 81–82,
116–117, 120, 131, 143
Dataverse, 77, 78
Digital Curation Centre (DCC), 77
Digital Public Library of America (DPLA),
77. *See also* eScience
Digital Science, 13
Discovery tools, 33, 73
Distributed Data Curation
Center (D2C2), 77
DOI, 143
Dryad, 75, 77
DSpace, 77–78
DuraSpace, 77–78

E

EBSCO, 34, 37, 61, 132
Editors, 20–21, 22, 23, 24, 25, 26, 29–30, 31,
103, 104, 105, 117
Electronic laboratory notebooks (ELNs),
5, 6, 23, 75, 85–94, 99, 145. *See also*
eScience
eLife Lens, 13
Elsevier, 35–36, 37, 46
EndNote, 5–6, 46, 47, 48–49, 50, 50*f*, 51*f*,
54*f*, 61, 138, 145. *See also* Bibliographic
management

Engineering Village, 34
eResearch, 99–100. *See also* eScience
eScience, 3, 5, 6, 58, 71–74, 75–82,
99–100, 146. *See also* Data; Electronic
laboratory notebooks (ELNs)
Essential Science Indicators, 37, 103, 104,
105–107, 115, 118
Ethical issues in science, 19–27, 30, 44–45,
58, 75–76, 80
Evaluating journals, 103, 104–111, 112, 117

F

Facebook, 4, 15, 129. *See also* Social media
Figshare, 13, 75, 78
Fourkas, John, interview with, 29–31
Fraudulent results, 21–22, 30, 86
F1000Research, 10, 13, 101
Funding agencies, 16, 25, 34, 76, 81, 82, 104

G

Garfield, Eugene, 1, 24, 103, 104, 105,
115–121, 124, 125
Garfield, Eugene, essays by, 1, 3, 117*f*, 119
Garfield, Eugene, interview with, 115–121
Google Scholar, 16, 34, 35, 38, 67, 107–108,
125, 126
Gordin, Michael, 4, 5, 19–20, 25
Graduate students, 27, 30, 50, 65, 66, 67, 68,
80, 92–93

H

High-impact journals, 11, 20, 22, 104
h-index, 6, 107, 120, 141
Horizon 2020, 13

I

Image manipulation, 22–23
Impactstory, 131
InChI, 98–99, 143
InCites, 131
Information literacy, 43–53, 58–62, 65, 66,
67–68, 69, 100–101, 145–146
Institute for Scientific Information (ISI), 1, 37,
103, 115, 116*f*, 118, 123–125, 126
Integration of resources, 33, 34–37, 40, 91,
92, 93
IPython, 13, 89

J

Journal Citation Reports (JCR), 1, 21, 37, 103, 104–105, 115, 118
Journal Impact Factor (IF), 21, 31, 103, 104, 105, 109, 115, 116–117, 118, 119, 120–121

L

LabArchives, 88, 89, 89*f*, 90, 90*f*, 91. *See also* Data; Electronic laboratory notebooks (ELNs); eScience
Lawlor, Bonnie, interview with, 123–126
Learning outcomes assessment, 48–53, 68
LibGuides, 44–45, 48, 49*f*, 50–53, 51*f*, 52*f*
Librarians, 3, 4, 5, 6, 7, 43–45, 53, 58, 79–80, 81, 99–101, 146
Librarians, new roles for, 7, 31, 44, 79–80, 93–94, 100
Libraries, 3, 5, 7, 43, 58, 67, 79–81, 82, 99–101, 126, 134, 146
Libraries, supporting eScience, 3, 5, 79–81, 82
Library instruction, 3, 43, 44–45, 48–49, 58, 80. *See also* Information literacy
Library Publishing Coalition, 13

M

MEDLINE, 34–35, 36, 37–38, 38*f*, 39*f*, 40, 100–101, 140
Mendeleev, Dmitrii, 4, 19–20, 98
Mendeley, 5–6, 37, 46, 47, 48, 66, 131, 132. *See also* Bibliographic management
Metadata, 45, 72, 73, 74, 76–77, 79–80, 81, 143. *See also* Data; eScience
Mullins, James L., 5, 79–80

N

National Center for Biotechnology Information (NCBI), 13, 14, 34, 35
National Institutes of Health (NIH), 13, 14, 26, 34, 118
National Library of Medicine (NLM), 34–35
National Science Foundation (NSF), 26, 71, 76, 78–79
NISO, 73, 78, 127, 128, 133

O

Office of Research Integrity (ORI), 26
Office of Science and Technology Policy (OSTP), 13, 77
Online Resource Center for Ethics Education in Science and Engineering (ORCEESE), 26
Open access, 4, 9, 11–12, 14, 16, 20–21, 101, 120–121, 138
Open Access Infrastructure for Research in Europe (OpenAIRE), 14
Open Access Scholarly Publishers Association (OASPA), 14
OpenDOAR, 78
ORCID (Open Researcher and Contributor ID), 7, 14, 16, 137, 141, 142, 143

P

PaperCritic, 132
Peerage of Science, 14
PeerJ, 14
Peer review, 6, 7, 9, 10, 11, 13, 14, 15, 16–17, 20, 21–22, 25, 31, 71, 74, 101, 120–121
Periodic Table of the Elements, 4, 5
Plagiarism, 5, 6, 22–23, 26, 30
PLoS Labs Open, 14
Plum Analytics, 128, 132–133
Post-publication review, 31
Predatory journals, 12, 66
Preprints, 12, 14, 40, 98–99, 108, 131
ProQuest, 34
PubChem, 34, 35, 60, 61, 125
Public Knowledge Project (PKP), 14
Public Library of Science (PLOS), 14, 126, 130–131
Publons, 14, 129
PubMed, 14, 34–35, 37, 38, 47, 49, 50*f*, 59, 61, 112, 143

R

Ranking of resources, 21, 33, 50*f*, 51*f*, 104, 111
R&D, 10
Reaxys, 4–5, 34, 35–36, 49, 50, 50*f*, 56*f*, 60, 61, 146
re3data, 77, 78

Research Data Alliance (RDA), 14
ResearcherID, 7, 16, 67, 138–140, 142, 143
ResearchGate, 16, 67, 133
Retraction Watch, 20, 22, 26
Retrieving information, 7, 20–21, 35, 37–38,
    40, 58, 72, 79–80, 115, 118
Reviewers issues, 11, 13, 14, 20, 21, 22, 25,
    26, 29–30, 31, 107, 120–121
ROARMAP, 14
Rubriq, 15

S

Schön, Jan Hendrick, 22, 25
Science Citation Index (SCI), 1, 37, 103,
    104, 112, 115, 116f, 118
Science courses, 58–62, 80, 91, 93, 145, 146
ScienceDirect, 34
Science librarians, 5, 6, 40, 79, 97–98
ScienceOpen, 15
Scientific communication, 9–17
Scientific fraud, 5, 6, 20, 22–24, 26, 88
Scientific fraud, detecting, 22–24
Scientific misconduct, 22–23, 25, 30
Scientific publishing, 3, 4, 9–17, 19–27,
    29–31, 44–45, 58, 67, 123–124
Scientific research, 5, 6, 13, 15, 71, 112, 134
Scientometrics, 65, 104, 115
SciFinder, 36, 37–38, 40, 49, 50–53, 50f, 51f,
    52f, 56f, 59, 60, 61, 100–101, 140, 146
SCImago Journal Rank (SJR), 111, 111f
Scopus, 37, 38, 108, 109, 112, 132, 141–142
Scopus Author Identifier, 141–142
SHared Access Research Ecosystem
    (SHARE), 12, 15
Shepard's Citations, 118
SHERPA/JULIET, 15, 78

SHERPA/RoMEO, 15, 78
SlideShare, 16
Social media, 9, 15–16, 97–98, 127–135
Social networks, 16, 40, 58, 67, 68, 134
SPARC, 14, 15
STM (Science, Technology, and Medicine),
    66, 68, 69
STEM (Science, Technology, Engineering, and
    Mathematics), 3, 4, 6, 9, 11–12, 16, 101
SurveyMonkey, 44–45, 48–49, 53, 54f, 58.
    See also Learning outcomes assessment

T

Thomson Reuters, 37, 46, 47, 103, 104–105,
    115, 131, 138–140
Twitter, 4, 15, 127, 129–130, 131. See also
    Social media

U

Unethical behavior, 6, 20, 22–23, 25, 26, 30
Unique identifiers, 7, 14, 137–142, 143

V

VIVO, 78, 91

W

Web of Science (WoS), 1, 6, 34, 37, 39f, 47,
    49, 50f, 59, 61, 66, 103, 115, 118, 119,
    120, 131, 138, 142, 143
Wiggins, Gary, interview with, 97–101

Z

Zenodo, 78
Zotero, 5–6, 46, 47, 48, 66. See also
    Bibliographic management